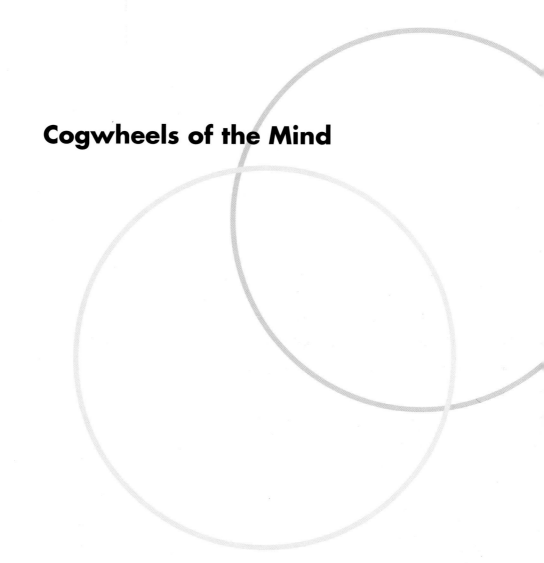

Cogwheels of the Mind

Cogwheels of the
MIND

The Story of Venn Diagrams

A. W. F. Edwards
Foreword by Ian Stewart

The Johns Hopkins University Press

Baltimore and London

Printed in the United States of America on acid-free paper

9 8 7 6 5 4 3 2 1

The Johns Hopkins University Press

2715 North Charles Street

Baltimore, Maryland 21218-4363

www.press.jhu.edu

Library of Congress Cataloging-in-Publication Data

Edwards, A. W. F. (Anthony William Fairbank), 1935–

    Cogwheels of the mind : the story of Venn diagrams / A. W. F. Edwards.

        p. cm.

Includes bibliographical references and index.

ISBN 0-8018-7434-3 (hardcover : alk. paper)

1. Venn diagrams. 2. Logic, Symbolic and mathematical. I. Title.

QA248.E28 2004

511.3'3—dc21

                              2003010633

A catalog record for this book is available from the British Library.

Frontispiece: Portrait of John Venn by C. E. Brock, 1899. By kind permission of the Master and Fellows of Gonville and Caius College, Cambridge.

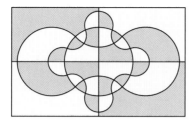

"Cogwheels of the mind," said Jules. "That would make a terrific title for a book."

"He thereby solved a problem that had remained unanswered for over a century," Lucretia concluded.

"Golly!" said Jules.

"It's one of those discoveries that are very hard to make, but very easy to belittle," I said. "It's not famous like the Poincaré Conjecture or the Riemann Hypothesis."

"Those are famous?" asked Jules.

"Among mathematicians. Venn diagrams aren't exactly the mainstream of mathematical research anymore. But, important or not, they picked up an intellectual loose end that really did need tidying up. And the answer is simple and elegant, the essence of mathematics. I think it's extremely clever. It grows on you."

"And it does *relate* to the mainstream," said Lucretia. "It has applications to combinatorics."

"Not only that!" I said. This was getting exciting. "In keeping with the times, the diagram is a fractal. I mean, the picture for infinitely many sets has structure on all scales, and is defined by a recursive procedure. Thus the Edwards–Venn diagrams are, you know, a visual icon for the modern era."

—Ian Stewart, *Another Fine Math You've Got Me Into . . .* , 1992

v

CONTENTS

For me, this book has deep roots.

I was unusually interested in mathematics from an early age, and by my teens I was greedily devouring every issue of *Scientific American,* mainly because it had a regular "Mathematical Games" column by the inimitable Martin Gardner. I probably became a mathematician because of that column: it brought the subject to life and showed me that new discoveries were constantly being made. For despite its recreational character, Gardner's column made frequent contact with the frontiers of mathematical research.

At the same time, I first came across Venn diagrams in a wonderful book by the equally inimitable Warwick (W. W.) Sawyer, *Prelude to Mathematics.* That's "prelude" as in the musical sense, not as "getting ready for." I learned about the strange rules of Boolean algebra, such as $1 + 1 = 0$—not bad for what (justly) claims to be the algebra of logic! And I was led to understand how simple graphical images can encompass a vast amount of insight and information.

In 1987 I inherited Martin Gardner's mantle, becoming the fourth person to write what by now had become the "Mathematical Recreations" column. I assume that, by this time, "recreations" were more politically correct than "games." More precisely, I started to write such a column for the magazine's French translation, *Pour La Science.* Not too many years later, I was promoted to the American parent magazine. But before that happened, I was following in Gardner's footsteps. And, like him, I got a lot of useful material from my readers. The column rapidly attracted quite a large mailbag, and while some of it was rubbish, a surprisingly large proportion was really good stuff, intriguing and provocative.

Some time in 1988—I regret I've lost the correspondence in some office clean-out— I received a package of material from Anthony Edwards, a Fellow of Gonville and Caius College, Cambridge. He had been asked to design a stained-glass window to commemorate the college's two most famous mathematicians, John Venn and Sir Ronald Aylmer (R. A.) Fisher. Fisher was a statistician, Venn a logician. As a result, Anthony had taken an interest in Venn diagrams. These are regions on a sheet of paper which represent sets; their overlaps represent the intersections of the corresponding sets, the members they have in common. For instance, if one set is "things that fly" and the other is "pigs" then the intersection is "flying pigs" like this:

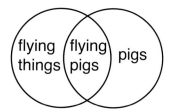

There are really four regions here: flying pigs; flying things that are not pigs, pigs that are not flying things, and things that neither are pigs nor fly. The first region is everything inside both circles, and, if the final region is puzzling you, it's everything *outside* the two circles.

It starts to get interesting when there are three sets; now you need three circles. (The book will fill you in on all this, so I won't go into the details. The point is that any combination of in and out for membership of the sets must correspond to a region.) With four sets, circles are no use! They didn't have enough possible combinations of overlaps. Venn himself found a way round this, using ellipses. But what of five, six . . . any number of sets? On this question, the great man was silent.

By using suitably convoluted shapes, it's easy enough to construct workable diagrams, but they tend to have one aesthetic flaw: some of the overlaps fall into several disconnected bits. Anthony had set himself the problem of devising elegant diagrams, for any number of sets—by which I mean that each overlap region is connected. And he'd solved it. His solution, in which successive regions had ever more wiggly boundaries, had a beautiful recursive structure: to increase the number of sets by one, just add another wiggly curve.

This was too good to resist, and I made it the subject of my fourteenth column. The setting for the column was a fictitious dinner at High Table in Gonville and Caius College. A mathematician and his guests were contemplating eight objects: a cat, a crow, a fur hat, an umbrella, a polar bear, a chameleon, a blonde wig, and a beach ball. What, they were asking themselves, did these objects have in common?

"You can get them all at Harrods," was the first attempt. But maybe not crows. Or polar bears. "You can't use any of them to travel to the Canary Islands." Better, but you might hitch a lift on a passing crow. Eventually, the mathematician put the guests out of their misery. "What they have in common is: they have absolutely nothing in common."

This enigmatic remark was then decoded. The objects have, or do not have, three distinct attributes: black, alive, hairy. The cat is black, alive, and hairy. The crow is black, alive, and non-hairy. The fur hat is black, dead, and hairy. The umbrella is black, dead, and non-hairy. The polar bear is non-black, alive, and hairy. The chameleon is non-black, alive, and non-hairy. The blonde wig is non-black, dead, and hairy. Finally, the beach-ball is non-black, dead, and non-hairy. Among them, these eight objects possess or do not possess all possible combinations of the attributes. That is, they define all possible overlaps of the three sets "black things, living things, hairy things."

Ah, but what about a fourth attribute—"harmonious"? The crow (possibly) has it, the cat does not . . . and so the column launches into the world of Venn diagrams, with Anthony's solution as the climax.

I needed a title. Looking at the shapes, wiggling in and out, I couldn't get my mind away from the image of cogs in a clock or a gearbox. And so I found my title, "Cogwheels of the Mind." And now here it is again, reincarnated in a far more glorious form.

Anthony's thinking has progressed considerably since those early days. He has added further conditions to the shapes he seeks—conditions like "symmetry." He has become a world expert on Venn diagrams. They may not be especially *useful*— but that's true of a lot of mathematics. There are more reasons for developing a mathematical idea than direct use. I doubt that many of you find a use for square roots in your daily lives, but I guarantee that our technological society would cease to function if all square roots were removed. The important thing about a mathematical idea is that it should contribute something of significance to the subject as a whole—which is undeniably useful. Every time you make a phone call, surf the Web, book a holiday, watch TV, or drive a car, you are relying on a vast array of mathematical concepts and techniques, all hidden away behind the scenes so that they don't frighten you.

It's wonderful when one of these ideas is hauled out to take center stage, and that's what Anthony does in this book. It may not be the most important piece of mathematics ever, but how often do you listen to music that is the most important piece of music ever, or read the most important novel ever? Believe me, if mathematicians could agree on what the most important piece of mathematics ever *was*, you wouldn't want to see it. It would be incomprehensible except to experts. So rejoice in a small but gleaming gem, the strange and intricate world of Venn diagrams. Give your mental cogwheels a spin. They'll love it, and so will you.

Ian Stewart
Coventry, July 2003

I started to write this book with the intention of providing a popular but accurate account of Venn diagrams from a geometrical rather than a logical point of view, with emphasis on the many recent and beautiful developments. But it has been an unruly child from the outset, and my own involvement in some of these developments has kept on intruding. This preface is both a warning about the nature of the book and a disclaimer that it is intended to secure priority of discovery for me or anyone else.

Mathematical discovery is perhaps the most delightful experience which academic life has to offer. The pure mathematician G. H. Hardy (1877–1947) wrote in *A Mathematician's Apology,* "It will be obvious by now that I am interested in mathematics only as a creative art," but Hardy was a mathematician's mathematician and most of us cannot appreciate his work. One of the joys of working with Venn diagrams is that there have been simple delights still to be uncovered that can be appreciated by the far wider audience of amateur mathematicians (amongst whom I count myself, for my Cambridge college, Trinity Hall, declined to admit me to read the mathematical tripos, for which I am grateful because it meant I became a scientist instead).

Hardy *created* beautiful mathematics, but working with Venn diagrams has been much more of a voyage of *discovery*. I cannot see a sense in which, to take an example from the book, the Venn diagrams with sevenfold rotational symmetry did not exist before they were found. We did not know they existed—Professor Branko Grünbaum, the authority on Venn diagrams, had even come to doubt the possibility—but the moment they were uncovered they seemed ageless and eternal. My discovery in December 1992 of the one I christened "Adelaide" (for that is where I found it) was a source of infinite pleasure to me that nothing can take away and is in no way diminished by the later knowledge that Grünbaum had found it earlier in the same year (but not published it). To savor the moment one needs only to believe at the time that one's discovery is new, and the more people who can share such an experience, the merrier.

The plan of the book is thus largely historical, whilst the mathematical level is mostly recreational, with many diagrams. Just as my earlier *Pascal's Arithmetical Triangle* (also published by the Johns Hopkins University Press) filled a rather surprising gap in the history of mathematics, so I hope that *Cogwheels of the Mind* will contribute to a knowledge of the history of Venn diagrams as well as providing a pleasurable addition to recreational mathematics.

I am indebted to Ian Stewart for the title and for contributing a foreword. I am grateful to Gonville and Caius College, Cambridge, for electing me into a fellowship many years ago, an act which not only led to an interest in John Venn and hence Venn diagrams but has also provided me with an intellectual and working environment second to none. Finally, it is a pleasure to acknowledge the freedom to study which my post in the Cambridge University Medical School has permitted.

CHAPTER 1

John Venn and
His Logic Diagram

Cambridge University is a place of long memories. John Venn, inventor of the diagram which bears his name, was born near Hull, on the east coast of England, on 4 August 1834 and entered Gonville and Caius College in the university as long ago as 1853, yet I have spoken to someone who remembered him.

The college was founded in 1348 by Edmund Gonville and enlarged and re-endowed in 1557 by John Keys, physician to Queen Mary. In adding his name to that of Gonville, Keys latinized it as "Caius," but the English pronunciation is always used. Just three hundred years later, in 1857, John Venn was amongst those who sat the mathematical tripos, the flagship university examination that tested the intense training of the previous three years. He was placed equal sixth in the overall order, the highest student of his college.

On the strength of this result Venn was elected into a college fellowship which he was to retain until his death, sixty-six years later. When I became a fellow of Caius in 1970 some of the older fellows still remembered him, and Sir Vincent Wigglesworth, the doyen of insect physiologists, told me how, as an undergraduate, he had one Sunday been accompanied to chapel by his mother and had observed Venn admiring her from his stall.

As a new fellow of the college I was able to enjoy the fine portrait of Venn in the college hall, and as a teacher of probability I became curious to know more about the man who had given his name to the simple but striking diagram so familiar to students of probability and logic, of discrete mathematics and computer science.

When my elder daughter, Ann, was fourteen she came home from school one day with the news that in mathematics she had been studying Venn diagrams, at the time a popular component of elementary mathematical teaching in Great Britain. "And did they tell you who Venn was?" I asked. "Oh, was he a person?" "Well, he *was*," I replied, "but he died a long time ago, in 1923, but since his house was in this parish we might be able to find him in the churchyard." And without difficulty we found his grave, and that of his wife and their only son, in the Trumpington extension churchyard beside the London road, quite close to our home.

In 1862 Venn became lecturer in Moral Sciences in Caius College, with responsibility for teaching the undergraduates what we now call philosophy, and especially logic. Indeed, the titles of his three major academic books all include the word logic—*The Logic of Chance,* 1866; *Symbolic Logic,* 1881; and *The Principles of Empirical or Inductive Logic,* 1889. From about 1869 Venn gave lectures in the "intercollegiate" series in which the colleges of the university joined forces, and he tells us in the preface to *Symbolic Logic* that these lectures formed the basis of the book, in which his logic diagram was described. In the course of developing his lectures he hit upon the idea of a *Venn diagram:*

> I began at once somewhat more steady work on the subjects and books which I should have to lecture on. I now first hit upon the diagrammatical device of representing propositions by inclusive and exclusive circles. Of course the device was not new then, but it was so obviously representative of the way in which any one, who approached the subject from the mathematical side, would attempt to visualise propositions, that it was forced upon me almost at once.

Venn invented his diagram to overcome the difficulties experienced in extending the logic diagrams of Leonhard Euler (1707–83) to complex logical problems.[1] An Euler diagram is a symbolic representation of a logical proposition such as "All *B* is *A*, some *A* is not *B*," which uses circles (or sometimes other shapes) to represent classes of things, as in Figure 1.1. In this example, the proposition would be true if *A* stood for the class "boats" and *B* for the class "sailing boats," for it would then mean "All sailing boats are boats, but some boats are not sailing boats." The diagram represents the obvious fact that no "*B*" is "not-*A*"—no sailing boat is not a boat as well—by virtue of there being no region corresponding to "*B* but not *A*."

When such pictorial representation is extended to more complex sets of logical propositions, however, difficulties set in quite quickly. For example, a logician faced with a series of propositions might wish to ascertain whether they were mutually consistent, and Euler diagrams do not provide a graphical "algorithm" for settling such a question. Venn's own description of the impasse, in 1880, is conclusive: ". . . we cannot readily

---

[1] This is the only footnote in the book. Notes, sources, and suggestions for further reading are found at the end of each chapter, and references at the end of the book.

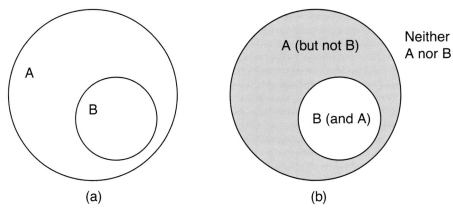

Figure 1.1. *(a)* An Euler diagram to indicate "All *B* is *A*, some *A* is not *B*." Note the convention that a label such as "*A*" refers to the entire area enclosed by the circle labeled. *(b)* A full description of the regions of the Euler diagram. A potential fourth region "*B* (but not *A*)" does not exist in this case, because there are no "*B*" which are not also "*A*."

break up a complicated problem into successive steps which can be taken independently. We have, in fact, to solve the problem first, by determining what are the actual mutual relations of the classes involved, and then to draw the circles to represent this final result; we cannot work step-by-step towards the conclusion by aid of our figures."

Venn had a better idea as he prepared to lecture to his Cambridge students. He did not wait until the publication of *Symbolic Logic* in 1881 to announce it, but wrote a separate paper, "On the Diagrammatic and Mechanical Representation of Propositions and Reasonings," published in July 1880, in which he suggested (after the passage already quoted) "a more hopeful scheme of diagrammatic representation":

Whereas the Eulerian plan endeavoured at once and directly to represent *propositions,* or relations of class terms to one another, we shall find it best to begin by representing only *classes,* and then proceed to modify these in some way so as to make them indicate what our propositions have to say. How, then, shall we represent all the subclasses which two or more class terms can produce? Bear in mind that what we have to indicate is the successive duplication of the number of subdivisions produced by the introduction of each successive term, and we shall see our way to a very important departure from the Eulerian conception. All that we have to do is to draw our figures, say circles, so that each successive one which we introduce shall intersect once, and once only, all the subdivisions already existing, and we then have what may be called a general framework indicating every possible combination producible by the given class terms.

The difference between Venn's diagram and Euler's is apparent even in the simple case we have so far considered. Figure 1.2 shows the Venn diagram corresponding to the Euler diagram of Figure 1.1. All four possible regions appear, namely "*A* and *B,*" "*A* but not *B,*" "*B* but not *A,*" and "Neither *A* nor *B,*" but since in our example "*B* but not *A*" does not exist, in the Venn diagram it is blacked out—the region is *empty.* With three classes *A, B,* and *C* there are of course eight regions to be depicted, and the corresponding basic diagram has the famous three-circle form with which Venn's name is so closely associated (Figure 1.3*a*; Figure 1.4).

In a footnote to his 1880 paper Venn throws a little more light on his thought processes. In 1854 George Boole, Professor of Mathematics in the newly established Queen's College in Cork, Ireland, had published *An Investigation of the Laws of Thought on Which Are Founded the Mathematical Theories of Logic and Probabilities*—or simply *The Laws of Thought* to modern readers—in which he advanced a revolutionary mathematical approach to logic. "It is a work of astonishing originality and power, and one which has only recently come to be properly

John Venn and His Logic Diagram

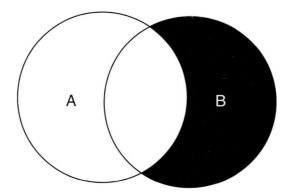

Figure 1.2. The Venn diagram corresponding to the Euler diagram of Figure 1.1. The region corresponding to "B (but not A)" is blacked out, indicating the nonexistence of this class.

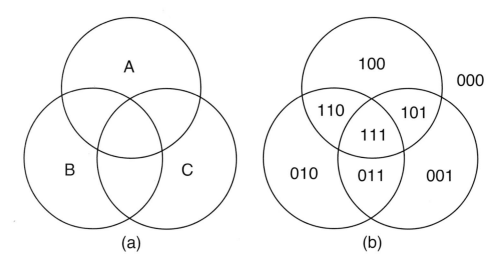

(a)

(b)

Figure 1.3. (a) The basic Venn diagram for three classes. (b) The eight regions labeled according to whether the thing considered is in class A (first digit), class B (second digit), or class C (third digit), "1" denoting presences and "0" absence.

Figure 1.4. The Venn diagram in the Hall of Gonville and Caius College, Cambridge, the work of Maria McClafferty. The lower window commemorates R. A. Fisher, statistician and geneticist (see chapter 3).

appreciated and to exercise its full influence on the course of logical speculation," opined the *Dictionary of National Biography* in 1886. The entry continued:

> Boole's work is not so much an attempt (as used to be commonly said) to "reduce logic to mathematics," as the employment of symbolic language and notation in a

wide generalisation of purely logical processes. His fundamental process is really that of continued dichotomy, or subdivision, in respect of all the class terms which enter into the system of propositions in question. . . . This process in its *à priori* form furnishes us with a complete set of possibilities, which, however, the conditions involved in the statement of the assigned propositions necessary [necessarily?] reduce to a limited number of actualities: Boole's system being essentially one for displaying the solution of the problem in the form of a complete enumeration of these actualities.

The author of this entry was in fact Venn himself, and it is remarkable that he was so modest as not to mention that it was he who had supplied the very diagram which made Boole's system of mathematical logic so accessible, Boole himself having suggested nothing in the way of diagrams. In his 1880 footnote Venn tells us, "I tried at first, as others have done, to represent the complicated propositions, there introduced [in Boole's method], by the old [Eulerian] plan; but the representations failed altogether to answer the desired purpose; and after some consideration I hit upon the plan here described."

In the present book we shall concern ourselves with Venn diagrams not as vehicles for mathematical logic so much as geometrical entities in their own right, and we shall not pursue the byways of logic further. Boolean logic is today normally represented through a *Boolean algebra* of 0s and 1s, and all we need notice for our purposes is that in a Venn diagram for n classes there are 2^n separate regions which can be put in one-to-one correspondence with the binary numbers 0 to $2^n - 1$, that is, with the n-digit numbers $000 \ldots 0$ to $111 \ldots 1$. Each digit corresponds to a class, 1 indicating that the thing considered belongs to that class, and 0 indicating that it does not.

In Venn's diagrammatic formulation each class is represented by a circle, the region inside being labeled 1 and the region outside 0. For $n = 3$ classes, the eight regions in the three-circle Venn diagram can thus be labeled as in Figure 1.3*b*. Each further class doubles the number of regions of course, just as the corresponding additional digit doubles the length of the list of numbers. What is not so obvious, though, is how this doubling of the number of regions is to be achieved when the new class is introduced into the Venn diagram with a curve "which shall intersect once, and once only, all the subdivisions already existing." From two congruent overlapping circles we easily proceeded to three, but, as Venn observed immediately, one cannot get four circles to intersect in all the ways required to deliver sixteen regions. What one *can* do as the number of classes increases is a principal theme of this book.

The abstract representation by binary numbers may seem to be an unnecessary burden for the reader to bear, but he can always slip back into reality by inventing an example to help him think concretely. Thus the classes *A, B,* and *C* of Figure 1.3 might be as mundane as the three courses of a meal—soup, main course, and dessert—in which case the binary numbers correspond to all the ways one can choose to have ("1") or not to have ("0") each course, from taking all courses (111) to declining everything (000). Aunt Tabitha is concerned about her weight and will leave out the dessert (110), but young Susanna, although she too is already watching her figure, still has a sweet tooth and opts for 011. Uncle Egbert's digestion, alas, cannot take the main course which is offered and he has to be satisfied with 101. And so on.

One famous three-class Venn diagram was drawn by Winston Churchill on 5 June 1948, whilst at dinner at Hever Castle, to indicate the mutual relations of the British Empire, United Europe, and the English-Speaking World, the United Kingdom corresponding to the intersection of these three classes (Figure 1.5).

Whilst one cannot draw a four-class diagram (or four-*set,* as we now more usually say) with four circles, Venn immediately came up with an elegant solution using four congruent ellipses (Figure 1.6), but he was of the opinion (erroneously,

John Venn and His Logic Diagram

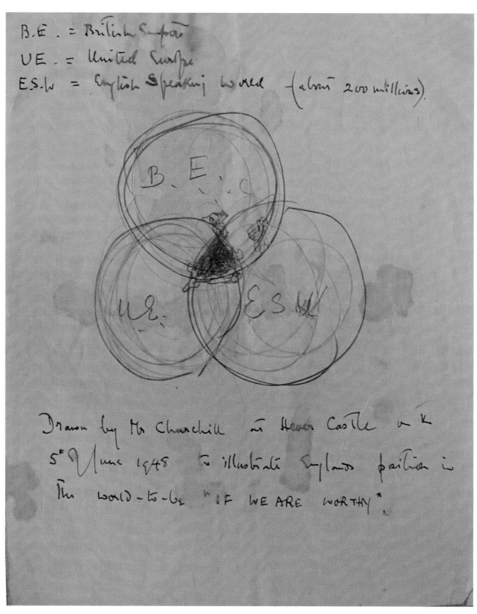

Figure 1.5. Winston Churchill's Venn diagram (1948). Reproduced by kind permission of the owner, Lady Dundas.

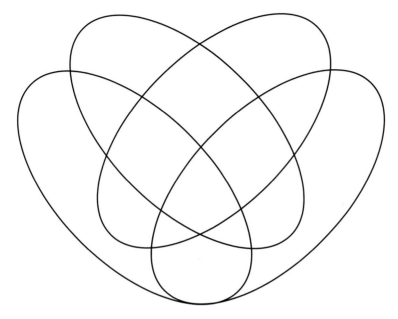

Figure 1.6. Venn's own diagram for four sets (1880).

as we shall see in chapter 7) that five congruent ellipses could not be formed into a Venn diagram.

John Venn was a noted historian as well as logician, and he looked carefully into possible precursors of his diagram. In addition to writing college and family history he perfected the genre of institutional history pursued through the lives of an institution's members. He published three volumes of the *Biographical History of Gonville and Caius College, 1349–1897* (five more have since been added), an example followed by universities, colleges, and schools all over the world. But nowhere were his historical skills more carefully displayed than in the final chapter of *Symbolic Logic*, entitled "Historic Notes," with its section II: "On the Employment of Geometrical Diagrams for the Sensible Representation of Logical Propositions." Venn was a man of great erudition, a major collector of books on logic. He gave his collection to Cambridge University Library in 1888, and the printed catalogue lists more than one thousand items. In the historical section of *Symbolic Logic* he reported on his researches into earlier diagrams.

Giving due credit to Euler and his near-contemporary Johann Heinrich Lambert (1728–77), Venn is careful to insist that the novelty of his diagram lies in its ability to represent *propositions* and not merely *classes* (sets). "The weak point about these [Eulerian circles] consists in the fact that they only illustrate in strictness the actual relations of classes to one another, rather than the imperfect knowledge of these relations which we may possess, or wish to convey, by means of the proposition. Accordingly they will not fit in with the propositions of common logic." He was obviously sensitive to the suggestion that other logicians had done almost as much as he had, for in the first edition of *Symbolic Logic* he writes that E. Schröder (*Operationskreis des Logikkalkuls,* 1877) and A. Macfarlane (*Algebra of Logic,* 1879) have shaded diagrams, but only to draw attention to the combination under consideration and not to express propositions. And later, in a footnote to "Historic Notes," he writes that both M. W. Drobisch (*Neue Darstellung der Logik,* 1836) and Schröder have used the three-circle diagram to represent combinations, but without having taken the additional step of using it as a basis for representing propositions. In the second edition of his book (1894) Venn adds a further reference in the footnote: "Dr J. Mich (*Grundriss der Logik,* 1871, p. 24) has come very near to this representation in one of his figures, where he employs three intersecting circles to represent class subdivisions, and then shades some of these. But though he uses these compartments to illustrate propositions, the shading itself has no propositional significance. The conception is still the Eulerian." Dr. Mich's figure is indeed a perfect three-set diagram (Figure 1.7), the sets referring to the presence or absence of qualities. Venn's own copy of Mich's book, which he bought in 1882, is part of the Venn collection in Cambridge University Library. In the next chapter we will encounter other

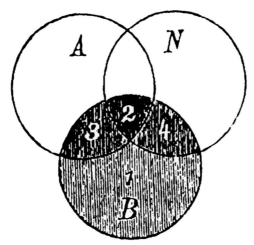

Figure 1.7. J. Mich's diagram (1871). Dr. M. Gratzke, of Gonville and Caius College, has kindly provided the following translation of the legend: "The relation between the notions of 'the pleasurable' (A), 'the useful' (N) and 'activity' (B) can be symbolized by three intersecting circles. And it is not hard to understand that these judgements follow from this relation: (1) Some activities are neither pleasurable nor useful; (2) Some activities are pleasurable as well as useful; (3) Some activities are pleasurable but all the same not useful; (4) Some activities are useful but all the same not pleasurable."

occurrences of the three-circle design, for it has long been used as a Christian icon, a heraldic device, and latterly a trademark.

Venn need not have been quite so sensitive. His diagram is now universally regarded primarily as a set or class diagram. It is not surprising that his extensive search through the literature turned up one or two examples of the three-circle diagram; in the history of ideas we can nearly always find such harbingers of great advances. Venn's own contribution, which fully justifies our attaching his name to the general diagram, was, first, to see that the diagram could and should be generalized to any number of sets (whatever its use); secondly, that it mapped Boolean algebra; and thirdly, that by writing *Symbolic Logic* he drew these advances to the notice of a wide public. Insofar as Boolean algebra supercedes the propositional logic, Venn's

John Venn and His Logic Diagram

claim to have invented the relevant diagram for propositions is fully justified, but the first interpretation of the phrase *Venn diagram* will now always be as a simple set diagram, and as such I will henceforth treat it.

The London, Edinburgh, and Dublin Philosophical Magazine and Journal of Science [Fifth Series], to give *Phil. Mag.* its full name, carried Venn's paper in its July 1880 issue. Claiming to be "Published the First Day of Every Month," the copy in Cambridge University Library is stamped 26 August, but already on 18 August, W. Stanley Jevons, Professor of Political Economy at London's University College, had written to Venn thanking him for a copy. "I have not yet had time to read the whole of it but I have already got the idea of your diagrammatic machine, of which I had previously heard some account from Mr Galton [Sir Francis Galton]. It is exceedingly ingenious and seems to represent the relations of four terms very well." Jevons is here referring to the final paragraphs of Venn's paper, where he reports having constructed a "logical-diagram machine" based on his four-ellipse diagram, in which the regions are made of wood and can be dropped by levers to correspond to the shading in the diagrammatic version.

Jevons was fascinated by such machines, having himself constructed a "Logical Abacus" and then a "Logical Machine" (shown in the frontispiece to his major work *The Principles of Science: A Treatise on Logic and Scientific Method,* first edition 1874), but he seems to have had a blind spot for diagrams. (In referring to Venn's paper in his letter he even omitted the word *Diagrammatic* from the title.) Like Venn, he appreciated the importance of representing Boolean algebra visually in some way, yet his letter does not even mention the benefits of the Venn diagram as such. He remarks in his *Treatise* that "these mechanical devices are not likely to possess much practical utility" but emphasizes their value for teaching and for promoting "the correct views of the fundamental principles of reasoning [which] have now been attained, although they were unknown to Aristotle and his followers. The time must come when the inevitable results of the admirable investigations of the late Dr. Boole must be recognized at their true value, and the plain and palpable form in which the machine presents those results will, I hope, hasten the time." But that was to be the function of Venn's diagrams, not Jevons's machines.

Venn must also have had a letter from H. J. S. Smith (1826–83), Savilian Professor of Geometry at Oxford, for the Venn papers in Gonville and Caius College contain, in addition to the letter from Jevons, a manuscript by Smith evidently prompted by Venn's diagram. (When Smith had been elected to the Savilian professorship in 1860, the other—somewhat tentative—candidate had been George Boole.) In this manuscript, which concerns the inclusion-exclusion theorem and its application to number theory, Smith suggests, "The curves may be drawn on any simply connected surface, or, instead of closed curve lines, we may consider closed surfaces in space. There is some convenience in regarding the classes as represented by curves drawn on a sphere, the surface of which represents the logical universe and is divided by any closed curve 'C' into two regions 'C' and 'not-C,' which we may term the inside and outside regions respectively."

As we shall see in chapter 3, this idea should have unlocked the solution to the problem of drawing Venn diagrams for arbitrary numbers of sets, but it lay fallow for more than a century until it independently resurfaced and inspired the general solution. It is as though Venn had no geometrical insight and Smith no practical interest in generalizing a Venn diagram.

NOTES

An excellent modern account of Euler's mathematical work is Dunham 1999, marred only by its omission of any discussion of Euler diagrams save in an unfortunate comment that perhaps Venn diagrams should be called Euler diagrams instead. There is as yet no full-length biography of John Venn, but notices of his life may be found in the appropriate places, such as the *Dictionary of National Biography (1922–1930)* (Weaver 1937) and the obituary notices of the Royal Society (Obituary of John Venn 1926). Venn's description of how he "hit upon the diagrammatical device" is from his *Annals 1834–1866* in the Venn manuscripts preserved in the Church Missionary Society Archive at Birmingham University.

In the Venn diagram in the Hall of Gonville and Caius College, Cambridge (Figure 1.4), the glass circles overlap each other so that the colors are true combinations. Professor John Mollon has drawn my attention to a letter to *Nature* in which the author (Allen 1871) described how he projected circles of red, blue, and green light so that they overlapped in precisely the same way (but the triple overlap producing white, of course). The date (1871) is interesting, for Venn was certainly a reader of *Nature.* Churchill's Venn diagram (Figure 1.5) must have been in his mind when, two years later, he said, "By our unique position in the world, Great Britain has an opportunity, if she is worthy of it, to play an important and possibly a decisive part in all the three larger groupings of the Western democracies. Let us make sure that we are worthy of it" (House of Commons, 26 June 1950; Gilbert 1988, 536).

The letter to Venn from W. Stanley Jevons and the draft paper by H. J. S. Smith are in the Venn papers, Gonville and Caius College. Smith is the subject of a chapter in *Oxford Figures* (Fauvel, Flood, and Wilson 2000), and there is a biography of George Boole by MacHale (1985). The standard historical account of logic machines and diagrams is in the second edition of the book by Martin Gardner (1983), and more recent information about the "Stanhope Demonstrator" described by Gardner is given by Weiss (1997). Venn's own account, "Historic Notes," in the second edition of *Symbolic Logic,* remains the fullest survey of the older material, especially of the more obscure variety. *Logic and Visual Information* by Hammer (1995) is a comprehensive survey which pays special attention to the difference between Euler diagrams and Venn diagrams, and to the "Peirce diagrams" of C. S. Peirce (1839–1914).

CHAPTER 2

Rings, Flags, and Balls

Not surprisingly, so simple a geometrical figure as the three-circle Venn diagram has surfaced in fields as varied as Christian iconography and industrial trademarks. In modern Cambridge alone one can see it on the notice board of Holy Trinity church and as the trademark of the Krupp company at a local airfield. Its oldest recorded use in Christian literature is in a thirteenth-century manuscript (alas destroyed by fire in 1944) from Chartres, France, in which the Trinity is represented by three interlocking rings (Figure 2.1). In the fifteenth century these became known as the "Borromean rings" through their incorporation into the crest of the Borromeo family of Northern Italy, and visitors to Milan and the Borromean Islands in Lake Maggiore may find countless examples of their heraldic use, not always interlocking in the same

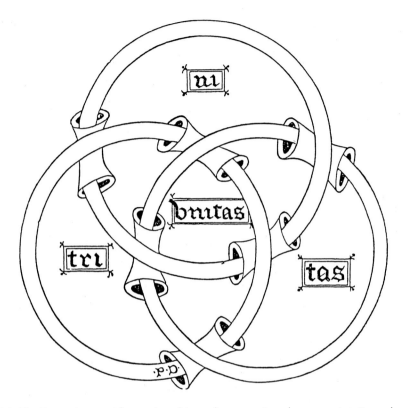

Figure 2.1. The Trinity *(tri-ni-tas)* from a late thirteenth-century French manuscript. Reproduced from Didron 1886, having originally appeared in Didron 1843.

pattern. In the Chartres diagram it is notable that the pattern is such that if any component ring of the Trinity is broken, the remaining two rings are disconnected.

As a part of mathematical knot-theory, the Borromean rings (though not under that name) first appeared in the work of P. G. Tait published in the *Transactions of the Royal Society of Edinburgh* (1877a). In addition to the interlocking pattern of the Chartres rings there are nine other possibilities, including the trivial one in which the three circles simply overlap. Such enumerations are the stuff of knot-theory, but of course the use of the pattern for a set-diagram ignores the three-dimensional nature of the Borromean symbol: Venn's circles are thoroughly welded together at the intersections.

Naturally there is no reason why a set in a Venn diagram should be represented by a circle, so long as the shape is a closed curve. Lewis Carroll (1832–98), otherwise known by his real name, Charles Dodgson, was Venn's opposite number at Christ Church, Oxford, where he was lecturer in Mathematics and had a different idea. He thought (Carroll 1896) that the universal set, or "universe," should be represented not by the whole plane as in Venn's diagram but by a closed area, and that rectangles rather than circles should be used (in fact, on p. 187 of the first edition of *Symbolic Logic,* Venn had already allowed the alternative of a closed area for the *Universe*):

> My Method of Diagrams *resembles* Mr. Venn's, in having separate Compartments assigned to the various Classes, and in marking these Compartments as *occupied* or as *empty;* but it *differs* from his Method, in assigning a *closed* area to the *Universe of Discourse,* so that the Class which, under Mr. Venn's liberal sway, has been ranging at will through Infinite Space, is suddenly dismayed to find itself "cabin'd, cribb'd, confined," in a limited Cell like any other Class! Also I use *rectilinear,* instead of *curvilinear,* Figures; . . .

Such a "Lewis Carroll diagram" starts with a square universe, which is then divided into two equal rectangles to represent the first set and its exterior. The boundary of the second set is placed similarly but at right angles, and the third set is then added as a central square (Figure 2.2). The equivalence to Venn's three-set diagram (plus an external boundary) is obvious, but Carroll then proposes a four-set diagram whose equivalence to Venn's requires a moment's reflection. Elongating the central square of the third set into a rectangle, he places a similar rectangle for

19

Rings, Flags, and Balls

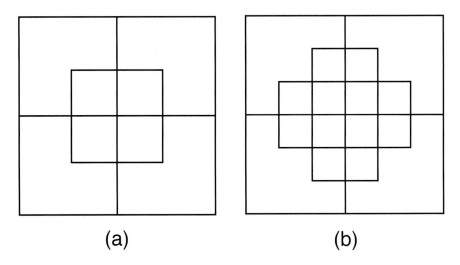

Figure 2.2. Lewis Carroll diagrams for *(a)* three sets and *(b)* four sets.

the fourth set at right angles to it. We shall see in the next chapter that when Carroll comes to a fifth set his courage fails him, for he simply divides each of the sixteen squares of the four-set diagram in two to get the thirty-two regions required, but of course the fifth set is then no longer represented by a closed curve. Had he persevered with his method he might have solved the problem of constructing a diagram for an arbitrary number of sets a century earlier than its eventual solution.

Carroll's diagram for three sets originally appeared in 1887 in a little book, *The Game of Logic.* The board and counters for the game came in an envelope accompanying it (Figure 2.3). Then, in 1896, Carroll published the first part of a planned three-part work *Symbolic Logic.* This part of the work was divided into five "Books," Book III being entitled "The Biliteral Diagram" and Book IV "The Triliteral

Diagram," the latter diagram being the same as that in the Game of Logic. Moreover, the board and counters for the Game of Logic were still available: "An envelope, containing two blank Diagrams (Biliteral and Triliteral) and Counters (4 Red and 5 Grey), may be had, from Messrs. Macmillan, for 3*d.*, by post 4*d.*" The author said that he hoped to publish further parts of *Symbolic Logic* "should life, and health, and opportunity, be granted to me," a Part II "Advanced" and a Part III "Transcendental," but he died early in 1898 without finishing them. He planned Part II to contain a treatment of "Multiliteral Propositions." In an "Appendix, Addressed to Teachers" in the published Part I (from which the above quotation is taken) he had given advance notice of the diagrams appropriate for up to ten sets "which will appear in Part II." Carroll must have changed his mind, for in 1977 W. W. Bartley III managed to reconstruct most of Part II from galley proofs and manuscript material scattered in libraries throughout the world (Carroll 1977), but found no further use made of the "multiliteral" diagrams.

It is a curious fact that if you draw an endless line on a piece of paper so that it cuts itself any number of times (but never cuts itself more than once at the same point), then you can color the resulting regions using only two colors without any adjoining regions being the same color. This seems to have been first noticed by Tait whilst thinking about knots, and he demonstrated it at the British Association meeting in 1876 (Figure 2.4). Venn diagrams also possess this property, but for a separate reason, which at first sight seems to be nicely demonstrated by induction. Suppose the property holds for a diagram with n sets. Then the closed curve for the next set added, the $(n + 1)$th, has to pass through each of the 2^n regions of the existing diagram. In the new diagram there will then be 2^n regions inside the new set and 2^n outside. Change the color of each of those *inside* the new set and the property now holds

21

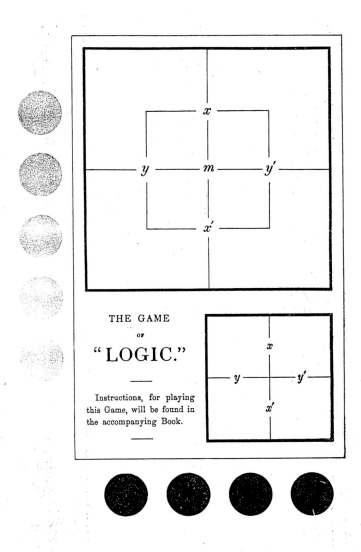

Figure 2.3. The board and counters for the Game of Logic, from Lewis Carroll's *The Game of Logic*.

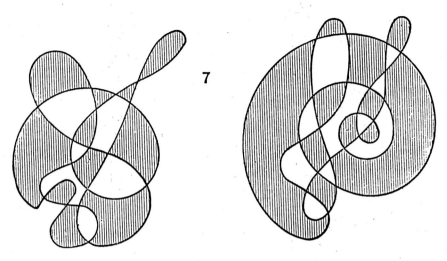

Figure 2.4. Closed plane curves are two-colorable. From Tait 1877b.

for a diagram with $n + 1$ sets, because all the regions which have been divided by the new curve are differently colored. But it certainly holds for a one-set diagram, so it must hold for any number of sets.

Mathematicians will raise two objections. First, we have not demonstrated that one can always add another set to a Venn diagram, and secondly, we have not demonstrated that all possible Venn diagrams can be thus constructed. The first point we ignore for the moment as being too advanced, but the second point we must address immediately, for in chapter 7 we will indeed meet a Venn diagram possessing the counterintuitive property that *none* of its sets can be removed in such a way that the remainder form a Venn diagram. In other words, it could not have been constructed by the processes we have so far discussed. Our proof therefore needs polishing, and binary numbers come to our aid. As we have already seen, a property of every n-set Venn diagram is that it maps the n-digit binary numbers from

23

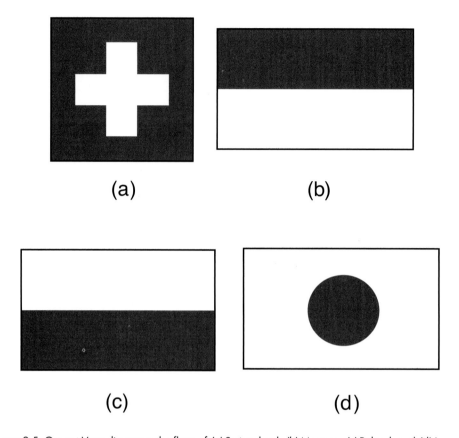

(a)

(b)

(c)

(d)

Figure 2.5. One-set Venn diagrams: the flags of *(a)* Switzerland, *(b)* Monaco, *(c)* Poland, and *(d)* Japan.

000 . . . 0 to 111 . . . 1. What is more, adjacent regions will necessarily differ by just a single digit, for hopping over a curve changes the digit corresponding to that curve's set. And we can color all the even numbers white and all the odd numbers black, thus proving our result.

Since every Venn diagram is thus "two-colorable" it is quite convenient to use this fact when drawing them in black and white, and we shall often do so in this book. No longer will a black region necessarily indicate the "emptiness" of the logicians.

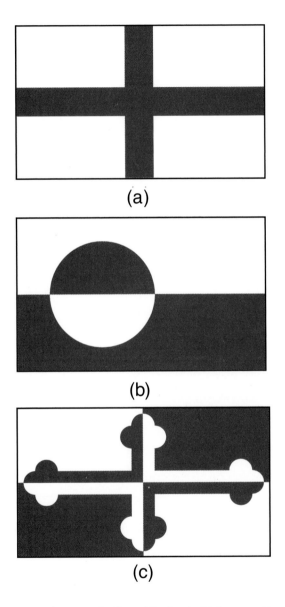

(a)

(b)

(c)

Figure 2.6. *(a)* Not a Venn diagram: the flag of England. *(b)* A two-set Venn diagram: the flag of Greenland. *(c)* A three-set Venn diagram: from the flag of the State of Maryland.

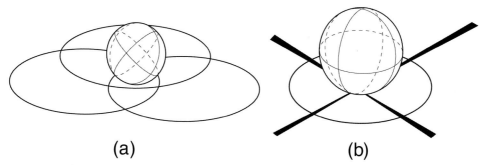

Figure 2.7. Plane (stereographic) projections of a three-set diagram on a sphere, leading to (a) the Venn and (b) the Carroll forms of diagram. These projections were first published in Edwards 1989.

We can now look around us for everyday examples of Venn diagrams, and flags provide a rich source with which to begin.

Even if we limit ourselves to national flags which are red and white, those of Switzerland, Monaco (identical to that of Indonesia), Poland, and Japan are all one-set Venn diagrams (Figure 2.5). By contrast, the flag of England (Figure 2.6a) is not, because the white areas of the field are disjoint. Passing on to two-set national flags, that of Greenland (Figure 2.6b) is interesting, because it shows clearly how to take a one-set flag, that of Poland, and add a set to it. (It can almost be obtained by adding a set to the Japanese flag, but the Greenland sun, unlike the Japanese one, is offset.) No country seems to have a two-color flag which can be thought of as a three-set Venn diagram, but heraldry offers wider scope. The flag of the State of Maryland consists of the arms of the first Baron Baltimore, George Calvert (1580?–1632), in which the first and third quarters are from the arms of his mother, Alice Crossland, daughter of John Crossland from Crossland in Yorkshire (Figure 2.6c). The heraldic description is "Argent and gules quarterly, over all a cross bottony countercharged," from which we learn that *countercharged* is the heraldic term for precisely the operation of coloring a new set added to an existing two-colored Venn diagram: always use the opposite color to that of the background. I look forward to hearing of a four-set heraldic Venn diagram, such as would appear if a central circle large enough to cut each arm of the cross were added to the Crossland pattern, and countercharged.

When we start looking for Venn diagrams on the surface of a sphere (as suggested by H. J. S. Smith and mentioned in chapter 1) a geographer's globe comes to mind. With only the line for the equator it is a one-set diagram; add the Greenwich meridian and it has two sets; continue with the Madison, Wisconsin, meridian, which is at about 90° west (taking it all round the earth along 90° east as well), and a three-set diagram appears, the globe having been partitioned into equal octants. When we project this diagram onto a plane using the geographer's stereographic projection (Figure 2.7) we recover either Venn's original form (if we project from the center of one of the octants) or a diagram with the same symmetries as Lewis Carroll's (if we project from one of the poles). As we shall see, the Carroll diagram holds the key to the canonical representation of an arbitrary number of sets.

Finally, note that a tennis ball is a one-set Venn diagram: one will come in handy in the next chapter. And take a look at a basketball as well.

NOTES

The information about the Borromean rings is from an article by Cromwell, Beltrami, and Rampichini (1998). At the time of writing this chapter the Cambridge University Library copy of Carroll's *The Game of Logic* still contained the board and the full complement of nine counters. Carroll's quotation "cabin'd, cribb'd, confined" is Shakespearean: *Macbeth,* act 3, scene 4. When I first tried to locate Carroll's *Symbolic Logic*

Rings, Flags, and Balls

28   (1896) in Cambridge University Library I failed because it was not in the main catalogue. It turned out that the library had put Carroll's children's books in the main catalogue but his academic books in its supplementary catalogue! Tait's British Association paper is in Tait 1877b. According to Wilson 2002, Tait also obtained the stronger result that if the number of lines meeting at each point is always even, then two-coloring is possible. This is, of course, the case with all Venn diagrams.

CHAPTER 3

Five and More Sets

So far we have seen Venn diagrams for one, two, three, and four sets, but have only hinted at the difficulties that lie ahead in drawing a diagram for five sets. Venn himself (1880) observed that it was in principle possible to include any number of sets: "for merely theoretical purposes the rule of formation would be very simple. It would merely be to begin by drawing any closed figure, and then proceed [*sic*] to draw others, subject to the one condition that each is to intersect once and once only all the existing subdivisions produced by those which had gone before." He gave four- and five-set examples (redrawn in Figure 3.1).

In the following year much the same material reappeared in *Symbolic Logic,* but Venn now added, "There is no need here to exhibit such figures, as they would

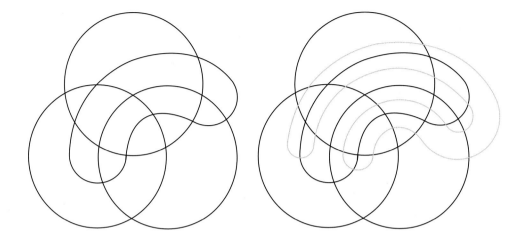

Figure 3.1. Four- and five-set diagrams, redrawn from Venn 1880.

probably be distasteful to any but the mathematician, and he would see his way to drawing them readily enough for himself," and, in a footnote,

> It will be found that . . . there is a tendency for the resultant outlines thus successively drawn to assume a comb-like shape after the first four or five. . . . The fifth-term figure will have two teeth, the sixth four, and so on. . . . There is no trouble in drawing such a diagram for any number of terms which our paper will find room for. But, as has already been repeatedly remarked, the visual aid for which mainly such diagrams exist is soon lost on such a path.

Figure 3.2 shows the attempt of the American logician C. S. Peirce (1839–1914) to persuade himself of the truth of Venn's statements, but not until More (1959) was

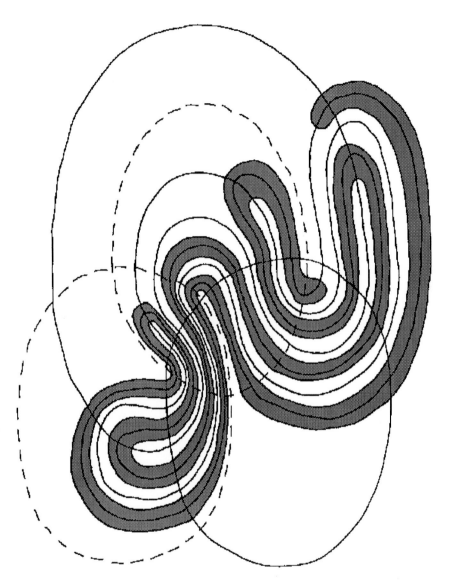

Figure 3.2. Corrected redrawing of C. S. Peirce's diagram, which carried the annotation "This shows that n closed curves can be drawn so as to separate the plane into just 2^n regions no more no less." Peirce had failed to make the sixth set boundary divide all the regions of his five-set diagram, so part of the boundary of the seventh set was wrong too. Redrawn from the original with the kind permission of the Department of Philosophy of Harvard University. My thanks to Professor Ivor Grattan-Guinness for bringing Peirce's drawing to my attention.

the matter settled formally. However, we must constantly remind ourselves that there may be—and, as we shall see, indeed are—perfectly good Venn diagrams which cannot be constructed by this process of addition.

Both Venn and Carroll gave up at four sets and offered five-set diagrams whose fifth set did not consist of a closed curve, so that some regions became disjoint. In our terminology, they were not really Venn diagrams at all: once one admits the possibility of sets being bounded by more than one closed curve, one might as well just list all the binary numbers between 0 and 2^n-1 and put a little ring round each! Venn's preference was for his four-set diagram with the fifth set added in the form of an annulus (Figure 3.3), but he noted (1880),

> Beyond five terms it hardly seems as if diagrams offered much substantial help; but then we do not often have occasion to meddle with problems of a purely logical kind which involve such intricacies. If we did have such occasion, viz. to visualize the sixty-four compounds yielded by the six terms X, Y, Z, W, V, U, the best plan would probably be to take two of the above five-term figures—one for the U part and the other for the not-U part of all the other combinations. This would yield the desired distinctive sixty-four subdivisions, but, of course, it to some extent loses the advantage of the coup d'oeil afforded by a single figure.

Venn went on, "We have endeavoured above to employ only symmetrical figures, such as should not merely be an aid to the sense of sight, but should also be to some extent elegant in themselves" (in *Symbolic Logic* he corrected this statement to "such as should not only be an aid to reasoning, through the sense of sight"). We shall shortly see how Venn's endeavor to employ symmetrical figures "elegant in themselves" need not have stopped at five sets.

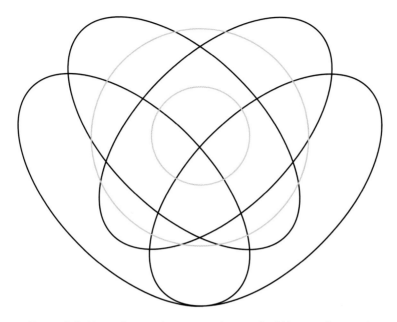

Figure 3.3. Venn's five-set diagram, redrawn. The fifth set is the annulus.

In 1896 Lewis Carroll tackled diagrams for five to ten sets in the appendix to his *Symbolic Logic*. As mentioned in the last chapter, for five sets he simply divided each region in his four-set diagram (see Figure 2.2*b*) with a diagonal line, rendering the fifth set disjoint. Here he has hit upon the idea that if each of the 2^n regions of an *n*-set diagram is filled with an *m*-set diagram the result will be an $(m + n)$-set diagram—though not one which is a Venn diagram in our sense, because of the disconnectedness of some sets. So for six sets he fills each of the regions of his four-set diagram with his square two-set diagram, and for seven sets fills each of the regions with his three-set diagram (see Figure 2.2*a*), and for eight sets fills each of the regions of his four-set diagram with the four-set diagram itself, replicating the figure in a kind of fractal manner (Figure 3.4). For nine sets he places two eight-set diagrams side by side (following Venn's plan), and for ten he uses four eight-set diagrams

Five and More Sets

arranged in a square. This still fits into his general pattern of replication, for he is essentially inserting an eight-set diagram into each region of a two-set diagram (as opposed to the alternative of inserting a two-set diagram into each region of an eight-set diagram).

It is all very Lewis-Carroll–like, but of course it is an admission of failure that he, like Venn, has not solved the problem in terms of true Venn diagrams. Ironically, as we shall now see, the solution to the general problem involves the symmetries that Carroll started with. But the story starts not with Carroll's symmetries so much as with an idea which, as we saw at the end of chapter 1, had occurred to another Oxford mathematician, Professor H. J. S. Smith. I shall tell the story just as it happened, and only then survey the literature to see how the simple solution could have been missed for so long.

The window depicting a Venn diagram in the Hall of Gonville and Caius College, Cambridge (see Figure 1.4), was put up in 1989, but the inspiration for it was not originally John Venn at all but the father of modern statistical theory, R. A. Fisher (1890–1962), whose centenary fell the following year. As Fisher's last undergraduate student I wanted the college to celebrate the centenary, and I set about persuading my colleagues to allow me to commission a stained-glass window based on the colorful design of a 7 x 7 Latin square which had appeared on the dust jacket of Fisher's pioneering book *The Design of Experiments* in 1935. I planned to install this in the lower light of one of the Hall windows, which was rectangular, but this raised the question of what to do with the upper light, whose top was the shape of a Gothic arch. It really needed another geometrical diagram, preferably curved at the top to fit the arch, and Venn's three-circle form suggested itself immediately. Nothing could be more appropriate, for both Venn and Fisher had been presidents of the college in their time.

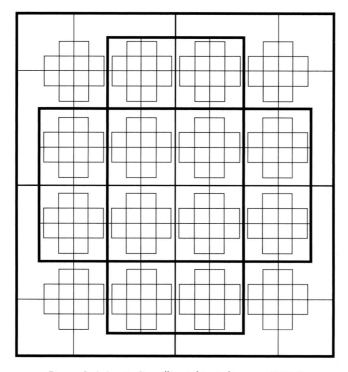

Figure 3.4. Lewis Carroll's eight-set diagram (1896).

In the course of persuading the Fellows of Caius to support this project I prepared some explanatory material in the summer of 1988, using a home computer to
draw the Latin square and the Venn diagram. It does not need a particularly inquir-
ing mind to ask, After the three-set diagram, how does one accommodate a fourth
set? And, having seen that four circles could not do the trick, I started drawing
ellipses, which fortunately my computer could manage. The pleasure of succeeding
with four similar ellipses (see Figure 1.6) was tempered with the feeling that I had
seen the diagram somewhere before, and indeed the next day the college's copy of
Symbolic Logic revealed where. Undeterred, I wondered about five sets, having found
Venn's solution unacceptable.

Five and More Sets

During my musings over the difficulty of devising a good five-set diagram I came to see that the awkwardness of Venn's three-circle form, from the point of view of generalization, probably lay in the asymmetrical way in which it treated the inside and the outside of each set. Representing a set by a circle in an infinite plane means that there is a gross difference between the region representing the presence of an attribute (the inside of the circle) and the region representing its absence (the whole of the rest of the infinite plane). Fortunately I did not then know that Lewis Carroll had complained about this (see chapter 2) and had solved the difficulty by giving the universal set its own boundary; indeed, nowadays one often sees Venn diagrams depicted using this convention. My own thought-process was to argue that we needed a finite but unbounded space in which to draw a diagram, and I knew enough mathematics to recall that the simplest space with this property was the surface of a sphere. In the twinkling of an eye I saw that the three-set diagram consisted of the equator and two meridians at 90° to each other (see Figure 2.7; I was fortunate to have been interested in astronomy and to have studied spherical geometry in high school). Each of the eight regions of the Venn diagram was now an identical octant of a sphere, and nothing could be more symmetrical.

A boundary for the fourth set instantly suggested itself: a wavy line crossing the equator four times, exactly in the shape of the seam of a tennis ball. I took a piece of scrap paper and some colored pens and eagerly sketched the scheme (Figure 3.5). The fifth set immediately demanded to be a wavy line crossing the equator eight times, and the sixth set a wavy line crossing it sixteen times, and the seventh set thirty-two times, and so on. The complete solution revealed itself in its utter obviousness, apparently without further intervention from me, as if my role was merely to peel the skin off an orange to reveal the structure of the inside.

I had another mathematical trick up my sleeve, for as an undergraduate I had

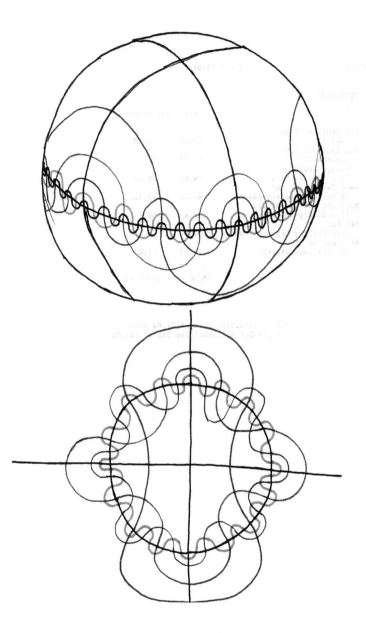

Figure 3.5. The original Edwards–Venn diagram.

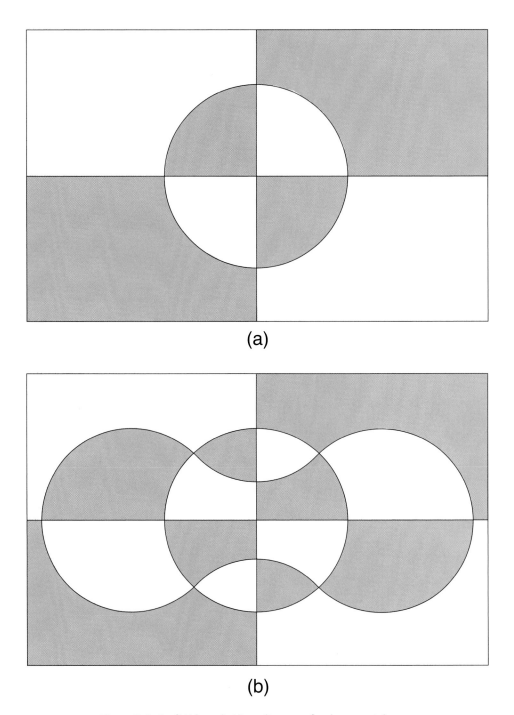

(a)

(b)

Figure 3.6. (a–f) Edwards–Venn diagrams for three to eight sets.

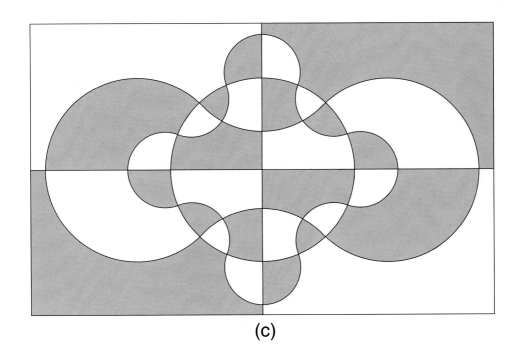

(c)

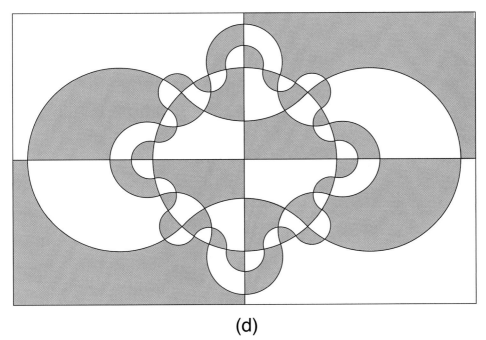

(d)

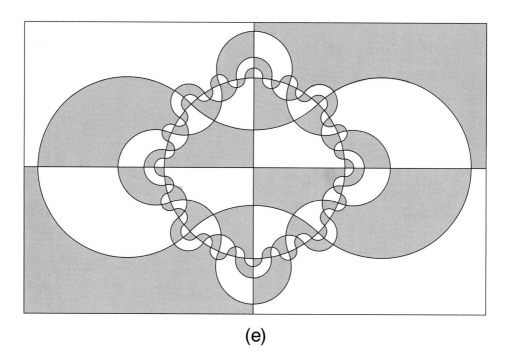

(e)

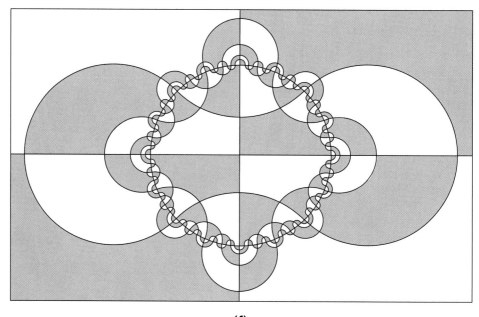

(f)

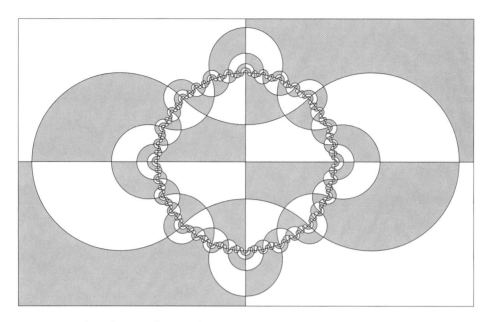

Figure 3.7. Edwards–Venn diagram for nine sets.

studied crystallography, amongst other things, so I knew about stereographic projections. Take a sphere and place it on a plane, just as a ball stands on a table. Call the point of contact the south pole. The stereographic projection is the one in which every point on the surface of the sphere is projected onto the plane by drawing the line from the north pole through the point and onto the plane. In particular, I knew that any circle on the sphere projected into a circle on the plane, so that if I represented the set boundaries after the first three by linked semicircles, then in the projection they would be linked arcs of circles. Nothing could be simpler, and I drew the resulting diagram, actually for seven sets, and dated it the last day of August 1988.

Ironically, the plane being infinite, I was now back to the type of unbounded Venn diagram to which Carroll had objected! I had kicked away the ladder by which I had ascended (the very ladder which H. J. S. Smith had erected a century earlier but had omitted to climb). The gains far outweigh the loss, however, for not only can we always meet the objection by going back to the surface of a sphere where these diagrams so naturally belong, but it is obvious from the general diagram (Figures 3.6 and 3.7) that this is *the* solution—there will never be a clearer representation of a Venn diagram for an arbitrary number of sets. We can see exactly how to add as many sets as we like, though of course the diagram becomes more complex, just as it must. It is like repeatedly folding a piece of paper in two—it soon becomes unmanageable, but that is just the way the world is. Ian Stewart, in an article in *Pour la Science* in 1989, called these diagrams "les dentelures d'Edwards–Venn"; his English-language version appeared in 1992 as "Cogwheels of the Mind" and is the source of my title for this book.

What is the trick? What precisely is the difference between my method of constructing a diagram in the plane and all the methods, starting with Venn's own, which have led to spaghetti-like figures that "would probably be distasteful to any but the mathematician"? The difference is simply that Venn and his successors added each new set boundary *relative to the last one added,* but I add each new set boundary *relative to the same original set boundary*—the "equator." So each time they added a set, Venn and his followers compounded the number of teeth of the "comb-like shape," but each time I add a set it has just the number of teeth, or "cogs," that it needs—2^{n-3} to be precise—and does not ride on the undulations of its predecessor.

A further irony is that the symmetry of my diagram for up to four sets is essentially the same as Carroll's (see Figure 2.2). It was only when he added a fifth set that his courage failed him. Indeed, when I published my solution in the *New Scientist* in January 1989, a reader—Michael Lockwood, appropriately from Oxford University—wrote in with a picture of what Carroll's seven-set diagram would have been. However, my curvilinear version has the undoubted advantage of making it easier for the eye to follow the boundary of each set without becoming confused where boundaries cross.

Let us now return to the tennis ball, whose seam so accurately portrays the shape of the fourth set. Take a ball and draw on it the equator in such a way that it cuts the seam in four equally spaced points (there is only one way you can do this!). Now mark the north and south poles. Next draw the Greenwich meridian, which we define to be the meridian which cuts the seam in its two points nearest the north pole. Finally, add the Madison meridian 90° away from it—which of course will be found to cut the seam at its two points nearest the south pole. You should now possess a tennis ball looking like the one I marked up on 31 August 1988 (Figure 3.8). A good alternative is a table-tennis ball, which is easily drawn on; it has an equatorial seam, so the fourth set will have to be added by hand.

Just how original are the Edwards–Venn diagrams? In a topological sense, not at all, because every n-set Venn diagram constructed by the sequential addition of sets has the same structure when considered as a mathematical graph. Very often in the history of mathematics hindsight reveals false starts which nevertheless contain the elements of a solution. H. J. S. Smith *might* have wondered how to draw more sets on a sphere and Lewis Carroll *might* have employed the symmetries of his diagram to greater advantage. A very remarkable advance was made by Branko Grünbaum of the University of Washington, Seattle, in 1975. He set out to prove that it was possible to construct a Venn diagram for any number of sets using only convex set boundaries. Using (unknowingly) the same initial symmetries as a Carroll diagram, Grünbaum succeeded in demonstrating the possibility by adding polygons with the number of sides given by the appropriate power of 2. His diagram may be roughly described as an Edwards–Venn diagram in which the "cogwheels" are replaced by polygons with as many edges as the cogwheels have teeth; the symmetries of both diagrams are the same.

43

Figure 3.8. A tennis ball marked as a four-set Venn diagram.

Naturally the visual effect in Grünbaum's diagram is more confused, because the polygons converge on each other more rapidly than the cogwheels do, but he was trying to prove a theorem, not to draw a clear diagram. He proved his theorem, but it was not a catalyst for a satisfactory general diagram (in his hands or anyone else's), and I myself did not know of it in August 1988. Neither did Mike Humphries, a London schoolmaster, who produced almost exactly the same solution to the same problem in 1987. In fact even earlier than Grünbaum's solution, Poythress and Sun (1972) had come close to it but using an extended version of what we have hitherto regarded as a Venn diagram—they permitted set boundaries to intersect multiply, that is, more than two at a time. As we shall see in the next chapter, permitting this generalization opens up whole new vistas of Venn diagrams.

A more profitable line of attack started with a note by Boyd (1985), who (also unknowingly) took Lewis Carroll's four-set diagram and added a central diamond-shaped fifth set. This inspired Fisher, Koh, and Grünbaum (1988) to add further polygonal sets, resulting in a solution with essentially the same structure as an Edwards–Venn diagram but without the latter's appealing shapes. These authors pointed out that their diagram could also be drawn on a sphere. I refer to other, less-successful, attempts to generalize a Venn diagram in the notes at the end of the chapter.

Toward the end of the 1980s the time was evidently ripe for solving the century-old problem of drawing Venn diagrams for arbitrary numbers of sets using only "symmetrical figures . . . to some extent elegant in themselves." Symmetry and elegance are not solely aesthetic virtues, however, for it is they that provide the "aid to reasoning" that Venn sought. And from the outset my solution revealed fundamental connections with the Gray code, with binomial coefficients, and with the "revolving-door algorithm" (Edwards 1989, Stewart 1989), to which we now turn. It also led to a whole new family of related diagrams, which we consider in chapter 5.

NOTES

A Latin square in n colors is an n x n array such that each color occurs exactly once in each row and once in each column. Latin squares were first studied by Euler and are important in the statistical theory of the design of experiments.

Gardner's book (1983), especially note 2 to chapter 2, contains a substantial list of earlier references to attempts at drawing Venn diagrams for arbitrary numbers of sets. I have consulted nearly all of these, and those which are not mentioned explicitly in the present book are mostly poor solutions, usually involving rectangular figures. An example of this genre, not recorded by Gardner, is Anderson and Cleaver 1965.

The reason why my diagrams were published in a popular weekly journal (the *New Scientist*) and not in a mathematical journal is that the British government had recently

Five and More Sets

46 imposed a "Research Assessment Exercise" on the universities, an element of which was the scoring of papers in refereed journals, the scores being credited to departments. As a protest at such an absurd scheme I deliberately sent my paper to the *New Scientist* so that it would not "count." Unanticipated benefits were quick publication, a huge circulation, color printing—and a fee! The diagrams were first exhibited on 19 and 20 October 1988 as a poster at a Royal Society of London meeting "Fractals in the Natural Sciences."

CHAPTER 4

The Gray Code, Binomial Coefficients, and the Revolving-Door Algorithm

In mathematics a good diagram is an invaluable aid to clear reasoning, whereas a bad one can seriously mislead. But it is not often that a diagram is good enough to suggest fresh avenues of inquiry altogether. The Edwards–Venn diagram immediately revealed a connection between Venn diagrams and the Gray code of communication theory, as well as providing a simple picture of the "revolving-door algorithm" of combinatorial theory (for listing all the ways of choosing r things from n different things).

We saw in chapter 1 that the 2^n separate regions of a Venn diagram can be put into one-to-one correspondence with the binary forms of the numbers 0 to $2^n - 1$, that is, with the n-digit numbers $000\ldots0$ to $111\ldots1$. Each digit corresponds to a set, 1 indicating the interior of the set, and 0 its exterior. When we add a further set to an n-set diagram we thread it through all the 2^n existing regions one after the other, thereby establishing an order of the regions and their corresponding binary numbers. Venn himself (1880) noticed that crossing the boundary between two adjacent regions (his "compartments") must correspond to changing just one of the binary digits (his "terms"), and this property must mean that the order of the binary numbers 0 to $2^n - 1$ established by adding the $(n + 1)$th set is such that each number differs from each of its neighbors in respect of just one of the n binary digits.

Such an order is known in communication theory as a *Gray code*. There seems to be some uncertainty as to just who Gray was, the choice being between the American telephone engineer Elisha Gray (1835–1901), coinventor of the telephone with Alexander Graham Bell (with whom he had a famous legal battle over the

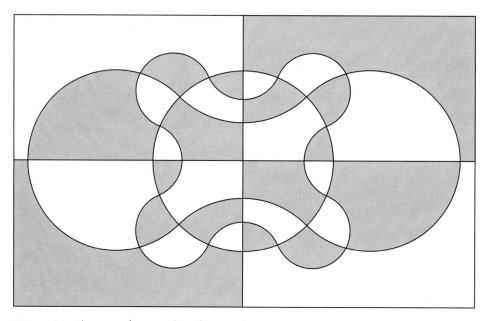

Figure 4.1. Alternative five-set Edwards–Venn diagram.

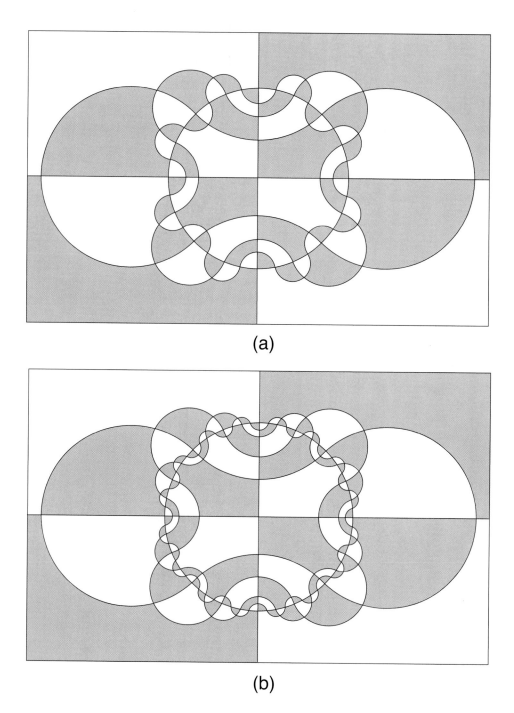

Figure 4.2. Alternative forms for *(a)* six-set and *(b)* seven-set diagrams.

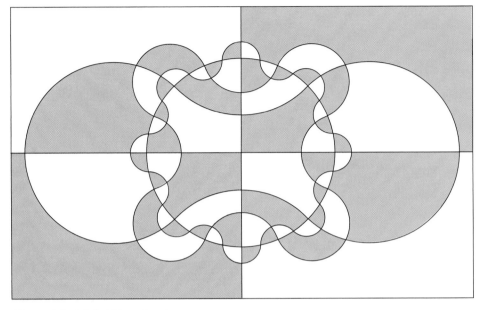

Figure 4.3. A hybrid form for six sets.

patent), and Frank Gray, who in 1953 filed U.S. patent 2,632,058 embodying the code. This is possibly no more than an example of Edwards's Law, that "every named law is attributed to the most famous person with that name."

In the days of electromechanical equipment it was advantageous to be able to move from one "number" to the next by changing only a single digit, for then there was no risk of generating an error through having to change two digits and not doing so exactly simultaneously. Nowadays the Gray code is more famous as a *combinatorial algorithm* which lists all the 2^n ways of making a selection amongst n different things (corresponding to the n sets) in an order which ensures that successive selections differ in only one of the things. This might be advantageous if some complex calculation had to be done for each selection, which could be shortened if one knew the result for a neighboring selection that differed from it in only one respect.

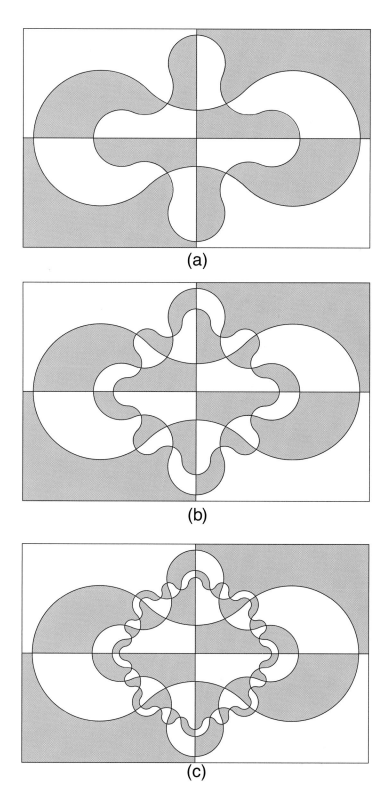

Figure 4.4. Edwards–Venn diagrams without the central circle, for *(a)* four, *(b)* five, and *(c)* six sets.

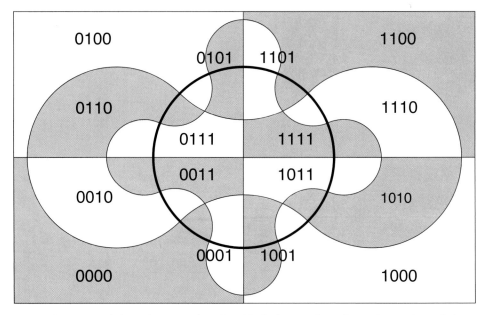

Figure 4.5. Figure 4.4a with a central circle added, showing how the circle cuts through the sixteen labeled regions in Gray-code order.

In an Edwards–Venn diagram the last set added is the one with the most "cogs," but it is simpler to think of the circle as being the last set (even though on the sphere it was the first—the equator!). This is because these two sets are in reality equivalent, for we could always gradually turn the cogged set into a circle provided we simultaneously allowed the circle to bulge in and out at the appropriate points to become a cogwheel. The effect is just the same as rotating the last cogged set by half the inter-cog angle (Figure 4.1). This leads to an alternative format for the diagrams (Figure 4.2), which is in fact the original format turned inside out and rotated through 90°. One may even generate hybrid formats (Figure 4.3). When the circular last set is removed the remainder still form a Venn diagram of course, also of quite attractive appearance (Figure 4.4).

Note that it is not true in general that the diagram remaining after a set is removed

```
0000
0001
1001
1000
1010
1011
1111
1110
1100
1101
0101
0100
0110
0111
0011
0010
```

Box 4.1

is a Venn diagram, as may be demonstrated in a striking manner by removing a set from Venn's own four-set diagram (see Figure 1.6). We find that there are two kinds of set, the kind that on being removed leaves a three-set diagram and the kind that leaves a diagram with disconnected regions, which is therefore not a Venn diagram.

Since the circle may always be removed, however, we are at liberty to regard it not as a set but simply as a line which passes through each of the regions, thereby defining a Gray-code ordering of them. Box 4.1 shows the four-digit code read from Figure 4.5. The code is of course circular, the listing starting at an arbitrary point. Every Edwards–Venn diagram generates a Gray code, but not all Gray codes can be represented by such a diagram.

To generate a Gray code we only need to observe that if we start with the one-digit code 0, 1 the following digit needs to be in turn 0 and 1 after the 0, and 1 and 0 after the 1. So we get 00, 01, 11, 10. This obviously ensures that all four two-digit numbers are created, but in addition it ensures that the adjacent numbers with the same first digit differ in their second digit, but adjacent numbers which differ in their first digit have the same second digit, exactly as required (Box 4.2). Moreover, repeating the algorithm repeats the property—always create a new last digit by adjoining 0, 1 and 1, 0 alternately to the existing numbers (Box 4.3). The final number always differs from the first number only in its first digit, ensuring the circularity of the code. (The midline will be used in a moment.)

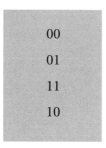

Box 4.2

We can see how the Edwards–Venn diagram grows in exact correspondence to a Gray code, for the pattern of 1s is simply the pattern of cogs belonging to the successive sets (digits). With each additional set the number of cogs is doubled up. Weave lines in and out of the digits, one for the first digit, one for the second, and

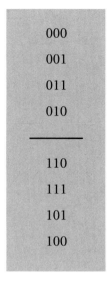

Box 4.3

one for the third, such that each line captures the corresponding 1s but not the 0s (Figure 4.6). When continued, this particular Gray-code construction generates the alternative-format diagrams shown in Figure 4.2.

Another simple algorithm for generating the above code is found by observing that the second half is the mirror image of the first half about the midline except for the first digit, which is always 0 in the first half and 1 in the second. Thus to add another digit we join to the bottom of the code a reverse-sequence copy of itself prefixed by 1s, at the same time prefixing 0s to the original code. Once again, the Gray-code property is preserved by the operation, in this case because the only point at which it might have been disturbed is at the mirror, but in fact it is ensured there too because the identical numbers on either side of the mirror gain different initial digits. And just as the first algorithm has the pictorial representation of adding a cogwheel, so this one can be pictured as a process of making a cut in the diagram from the center to the north

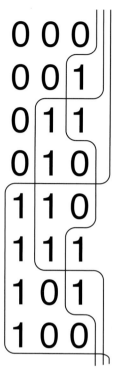

Figure 4.6. A three-digit Gray code showing boundary lines "capturing" the 1s in each column.

point, collapsing it round like a fan until the cut edges are pointing west and east, and then reflecting it about the east-west line. It all comes to the same thing!

It is easy to demonstrate that only one four-set Venn diagram arises from adding a set to the three-set diagram; we have met it in Venn's form (see Figure 1.6), Carroll's form (Figure 2.2b), and mine (Figure 3.6b), which are of course all the same thing. (For a little longer we continue to confine our attention to diagrams in which no three set boundaries intersect at a point, the so-called *simple* diagrams.) But how many ways can a fifth set be added to a four-set diagram? Even this, the easiest of the

Venn diagram enumeration problems, turns out to be quite complicated. Hamburger and Pippert (1997) found just twelve ways, or eleven if one were to work on the surface of a sphere. Each of these twelve will generate a Gray code. Nor is this all the five-set Venn diagrams that are possible, for in chapter 7 we will encounter a particularly beautiful one which cannot be constructed by adding a set to a four-set diagram (and is therefore said to be *irreducible*).

We saw how the Gray code led to a combinatorial algorithm that lists all the 2^n ways of making a selection amongst n different things in an order which ensures that successive selections differ in only one of the things. But what if we wanted a listing not of all the selections but only of those in which exactly r of the n things are selected? There are $\binom{n}{r}$ of these and, taking our cue from the Gray-code ordering, it would be convenient if we could order them in such a way that each differed from its predecessor in respect of exactly *two* of the things. Then we could proceed through the list by a succession of exchanges.

Consider the six-set Edwards–Venn diagram of Figure 4.7. Each region has been colored in accordance with the number r of the sets for which that region is part of the interior, or, in binary notation, with the number of 1s in the corresponding binary number. Thus the dark violet region corresponds to 111111, or the selection of all the things. This region must be surrounded by all the ways of leaving out just one of the things, for to cross any of its boundary lines is to exclude just one thing. Indeed, it must be a six-sided region surrounded by the six regions that correspond to selecting just five things. (The precise order will depend upon how we have chosen to match the sets to the digits.)

On inspection we see that these six regions form a necklace of blue "beads" joined corner to corner. Traversing this necklace we encounter all six selections

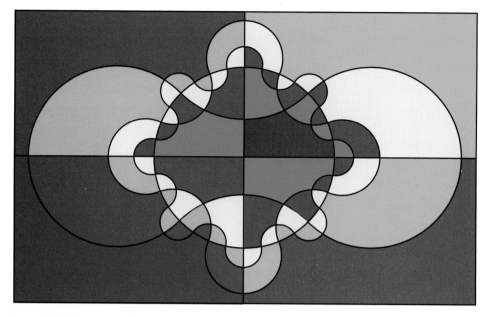

Figure 4.7. The regions of a six-set Edwards–Venn diagram colored according to the number of 1s in the corresponding binary number, revealing the operation of the revolving-door algorithm.

in just such an order as we require, for at each corner we cross two set boundaries simultaneously, leaving one set and entering the other. An exchange has taken place at the crossing point, which we can visualize as a revolving door through which two people pass simultaneously, one going in and the other going out, leaving the same number of people in each room. The combinatorial algorithm is indeed known as the "revolving-door algorithm," and the Edwards–Venn diagram makes it very real.

Outside the necklace of six beads corresponding to selecting just five of the things, there must be, by the same argument, a necklace of $\binom{6}{4} = 15$ beads listing the ways of choosing just four of the things, and indeed there is—colored green in Figure 4.7. The sequence continues through successive necklaces, with numbers of beads given by the numbers of the 6-row of Pascal's triangle, ending up with a single bead for 000000, the selection of nothing. One might begin to wonder why every bead does not have six sides, since there must be six numbers differing from "its" number by just one digit. The mystery is explained in chapter 6. Finally, we note from the diagram that each ordering of the selections of r things is itself embedded in the Gray-code ordering defined by the particular diagram.

A minor observation is that adding another set to a diagram gives a pictorial representation of the addition rule of binomial coefficients, best displayed by an example. Consider a four-set diagram. It has 1, 4, 6, 4, 1 regions, corresponding to the presence of zero, one, two, three, and four 1s, respectively, in their binary numbers. Adding the fifth set divides each of these regions into a part interior to the set, which adds one to the number of 1s, and a part exterior to the set, which adds only a 0. As Table 4.1 indicates, we end up with the next row of Pascal's triangle.

Although the identification of these combinatorial algorithms with a Venn diagram is new, the connection between Venn diagrams and Pascal's triangle was implicit in the nineteenth-century work of Jevons, mentioned in chapter 1. In 1870 he had correctly perceived the importance of enumerating all the combinations of sets, using capital and lower-case letters for the purpose: "These series of combinations appear to hold a position in logical science at least as important as that of the multiplication table in arithmetic or the coefficients of the binomial theorem in the higher parts of mathematics. I propose to call any such complete series of combinations a Logical Abecedarium." Then in his book (1877): "There exists a close connection between the arithmetical triangle described in the last section, and the series of combinations of letters called the Logical Alphabet." So when, in 1880, Jevons read Venn's paper, the idea that a Venn diagram is a sort of map of the binomial coefficients might easily have occurred to him, even if he never stated it explicitly.

Table 4.1.

The addition rule of binomial coefficients

Number of "1s"	0	1	2	3	4	5
Regions with this number						
Four-set	1	4	6	4	1	–
Added set	–	1	4	6	4	1
Five-set	1	5	10	10	5	1

NOTES

The correspondence between a Venn diagram and the Gray code and the revolving-door algorithm was first noted in Edwards (1989). I enunciated the law that "every named law is attributed to the most famous person with that name" (Edwards 1993) after having failed to persuade the editor of the scientific journal *Nature* to print a correction to a report which had attributed a result of Nicholas Bernoulli to his uncle James Bernoulli.

My source book for information about combinatorial algorithms is Nijenhuis and Wilf (1978). Readers familiar with a little graph theory will recognize the problem of adding a further set to a Venn diagram as equivalent to finding a Hamilton circuit on the graph of the diagram (Winkler 1984). The standard work on Pascal's arithmetical triangle is my book of 1987 (second edition, 2002), though I was ignorant of Jevons's extensive treatment of the subject (1877) when I wrote it.

The manner of labeling the regions of a Venn diagram with binary numbers contains an arbitrary element by virtue of both the order in which the sets are taken and the designation of the inside and outside of each set. I have chosen the labeling which, in chapter 6, will be found to correlate with the axis-convention of coordinate geometry, but this is not necessarily a good labeling for other purposes. Consistency requires, however, the adoption of one convention throughout.

CHAPTER 5

Cosine Curves
and Sine Curves

In September 1988 I circulated a draft paper, "Venn Diagrams for Arbitrary Numbers of Sets," to a dozen or so mathematical friends, asking them whether they had ever seen anything like my version of a Venn diagram. I could hardly believe that so simple a development had not been made many years previously. I remarked in the draft that as an alternative to the polar stereographic projection of the sphere which I had employed, "a Mercator projection would also be acceptable, though somewhat less elegant." I received a reply from Professor C. A. B. Smith, Weldon Professor of Biometry at University College London, well-known for his broad mathematical interests, including graph theory. He made the observation that in a Mercator style of diagram the sets could be represented by the family of cosine curves $y = \cos (2^{n-2}x)/2^{n-2}$, $0 \leq x \leq \pi$, $n = 2, 3, 4, \ldots$ (as in

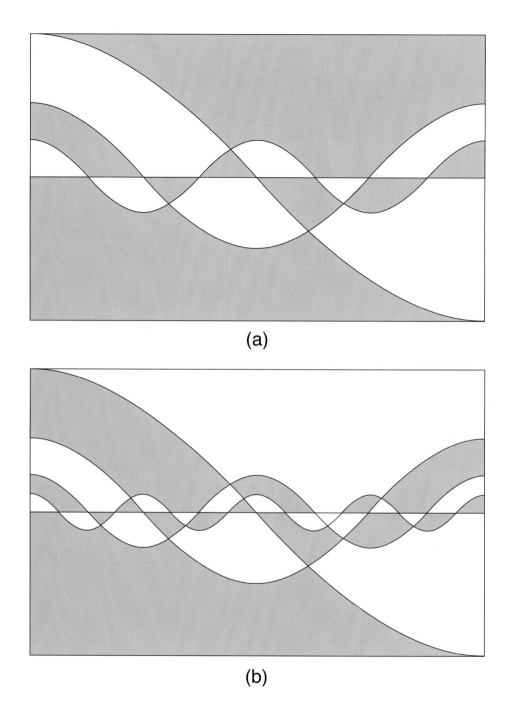

(a)

(b)

Figure 5.1. Cosine-curve diagrams for *(a)* four and *(b)* five sets.

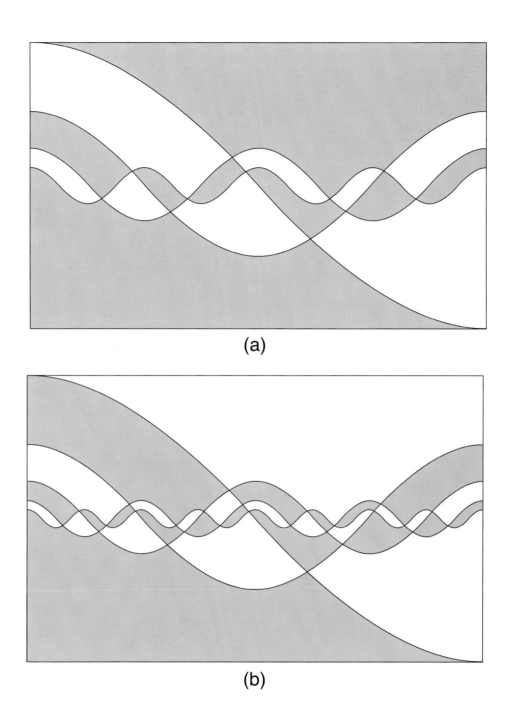

Figure 5.2. Cosine-curve diagrams without the equator for (a) four and (b) five sets.

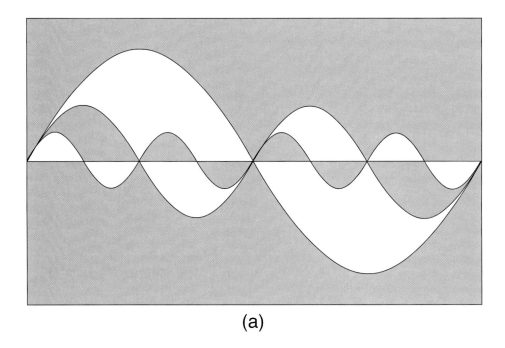

(a)

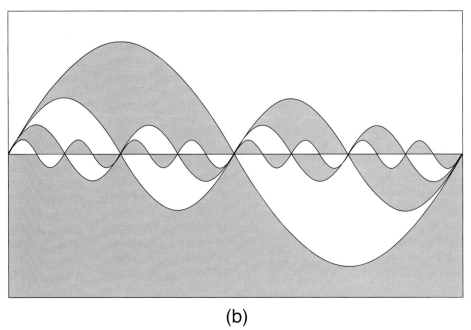

(b)

Figure 5.3. Sine-curve diagrams for *(a)* four and *(b)* five sets.

Figure 5.1). Here n is the number of the set, if we start with an equator $y = 0$ for the first set. We can also omit the equator, as before, leading to the curves of Figure 5.2.

However, Professor Smith's next observation led into altogether new territory: the corresponding family of sine curves $y = \sin(2^{n-2}x)/2^{n-2}$, $0 \leq x \leq 2\pi$, $n = 2, 3, 4,$ ... not only generated a Venn diagram (Figure 5.3) but, as we shall see, one that mapped the binary numbers in their natural order just as the cosine version maps a Gray code. In his letter Professor Smith did not actually draw any diagrams but relied on the trigonometric exposition, leading to the observation that there was an intimate connection between the cosine functions and the Gray code, and between the sine functions and binary counting. These connections were already familiar to electronics engineers; what is novel is the connection with Venn diagrams. A similar statement can be made about Fourier transforms: briefly, an even step-function generates the cosine curves and an odd step-function the sine curves.

We must specify in what sense the diagrams in Figure 5.3 are Venn diagrams, for hitherto we have not allowed set boundaries to intersect more than two at a time—a type of diagram known as *simple*—but here we find intersections of three or more boundaries. Yet the new diagrams are perfectly good Venn diagrams in that they divide the universe into regions as before, one for each of the numbers 0 to 2^{n-1}, and every set is enclosed by a single curve. We should certainly admit them to Venn's family. As before, we can omit the equator if we wish (Figure 5.4). Because of the intersections of several set boundaries at once, the identity of the boundaries on either side of an intersection is not explicitly maintained, but a moment's thought will always dispel any uncertainty.

Curiously enough, Martin Gardner once published a version of Figure 5.3*a* in

Cosine Curves and Sine Curves

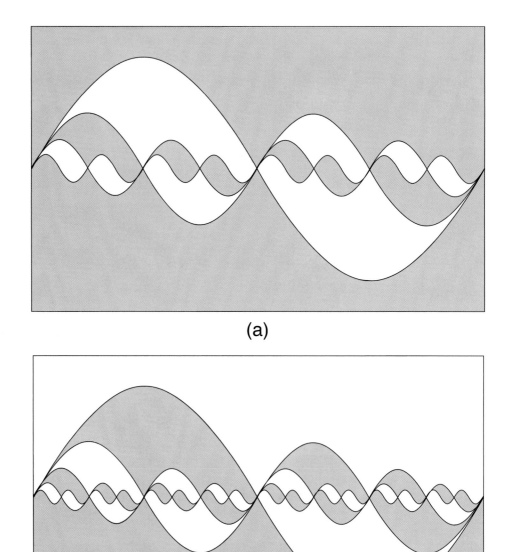

(a)

(b)

Figure 5.4. Sine-curve diagrams without the equator for *(a)* four and *(b)* five sets.

his "Mathematical Recreations" column in *Scientific American* (April 1971) whilst describing geometrical fallacies (redrawn rotated through 90° in Figure 5.5). The accompanying text read:

Theorem 4: Pi equals 2.

The ... illustration ... is based on the familiar yin-yang symbol of the Orient. Let diameter *AB* equal 2. Since a circle's circumference is its diameter times pi, the largest semicircle, from *A* to *B*, has a length of $2\pi/2 = \pi$. The next-smallest semicircles, which form the wavy line that divides the yin from the yang, are each equal to $\pi/2$ and so their total length is pi. In similar fashion the sum of the four next-smallest semicircles (each $\pi/4$) also is pi, and the sum of the eight next-smallest semicircles (each $\pi/8$) also is pi. This can be continued endlessly. The semicircles grow smaller and more numerous, but they always add to pi. Clearly the wavy line approaches diameter *AB* as a limit. Assume that the construction is carried out an infinite number of times. The wavy line must always retain a length of pi, yet when the radii of the semicircles reach their limit of zero, they coincide with diameter *AB*, which has a length of 2. Consequently pi equals 2.

That may be so, but how extraordinary that seventeen years before Professor Smith's sine-curve Venn diagram was first drawn, not only did it appear in a well-known column of "Mathematical Recreations," but the accompanying description envisaged proceeding to the "fractal" limit. It is perhaps surprising that none of the thousands of readers of the column remarked that Gardner had discovered a Venn diagram for an arbitrary number of sets. We were all asleep!

When demonstrating the properties of Professor Smith's "binary number" diagram I shall immediately revert to a polar projection of the version on a sphere, partly because it leads to very attractive pictures, and partly because there is then a

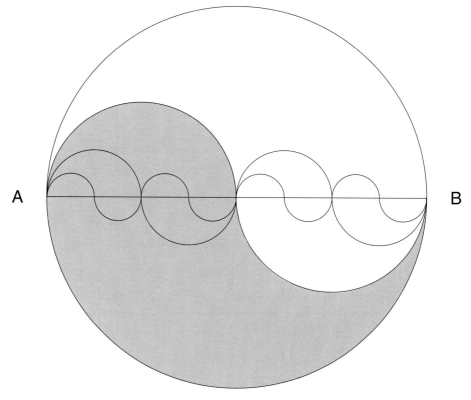

Figure 5.5. Martin Gardner's yin-yang diagram from 1971, not then recognized as a Venn diagram.

simple graphical equivalent to progressing from cosine curves to sine curves: a phase shift achieved by rotation. Figure 5.6 shows the binary form for three to six sets. It takes a moment or two to identify the individual sets, but each is exactly the same cogwheel shape as before. The two-set form is the flag of Greenland shown in Figure 2.6*b*. Once again, we can leave out the central circle and generate some more pretty diagrams (Figure 5.7).

The "binary" nature of these diagrams is easy to discern. Figure 5.8 shows the regions of the three-set figure without a central circle. The circle is once again not itself a set but a guide round the regions, passing through them in binary order. The last digit runs 0, 1, 0, 1, 0, 1, ... as the circle crosses in and out of the four-cogged wheel of the third set, and the middle digit runs 0, 0, 1, 1, 0, 0, ... as the circle crosses in and out of the two-cogged wheel of the second set. The first digit changes only as the circle crosses the straight boundary of the first set. The addition of further digits would correspond to the addition of further cogwheels, each with twice as many cogs as its predecessor. A striking feature of the diagram is its representation of the "carry" operation of counting in binary. For example, the next number after 011 is 100; adding 1 to 011 has had the effect of replacing the last digit by 0 and carrying 1 to the preceding digit, causing it in turn to be replaced by 0 with a further carry to the first digit, which becomes 1. All this is represented in the diagram by a triple boundary intersection—progressing from one number to the next involves crossing three boundaries simultaneously, out of two of the sets and into a third. Just as we did with the Gray code in the last chapter (see Figure 4.6), we can generate the linear form of the binary Venn diagram by adding curves to a simple listing of the binary numbers.

It is a surprising feature of Venn diagrams that the original simple forms turn out to correspond to the rather esoteric Gray code, whilst the more esoteric "sine-curve" forms with multiple intersections correspond to the common-or-garden binary code.

The three-set binary-from-Venn diagram can be drawn in ways different to Figure 5.6a. It certainly surprises students if one offers them a diagram looking like Figure 5.9 as a Venn diagram for three sets. It is easy to demonstrate that there are no other three-set diagrams (not counting as distinct any obtained by set boundaries merely touching without crossing). But as soon as one contemplates four sets, enu-

Cosine Curves and Sine Curves

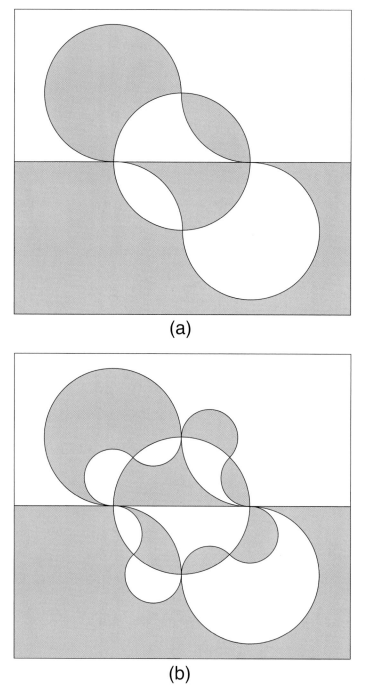

(a)

(b)

Figure 5.6. Binary-form diagrams for *(a)* three, *(b)* four, *(c)* five, and *(d)* six sets (Edwards and Smith 1989).

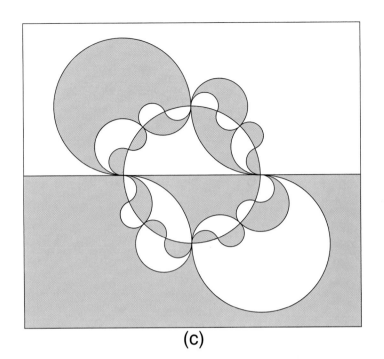

(c)

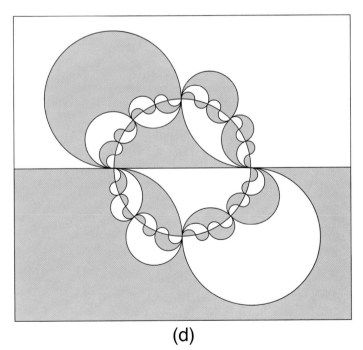

(d)

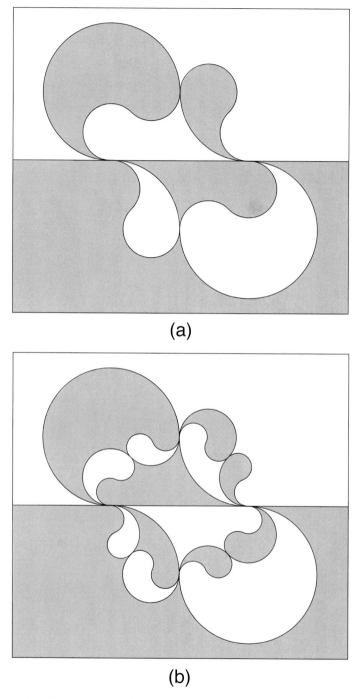

(a)

(b)

Figure 5.7. Binary-form diagrams without the central circle for *(a)* three, *(b)* four, *(c)* five, and *(d)* six sets.

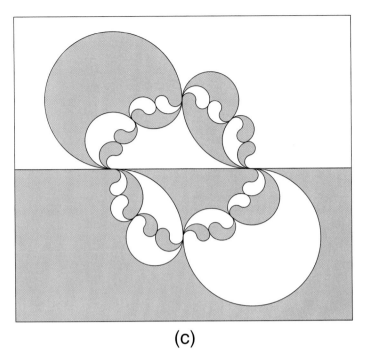

(c)

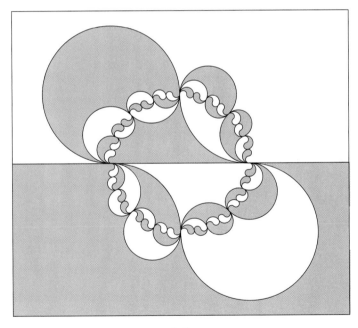

(d)

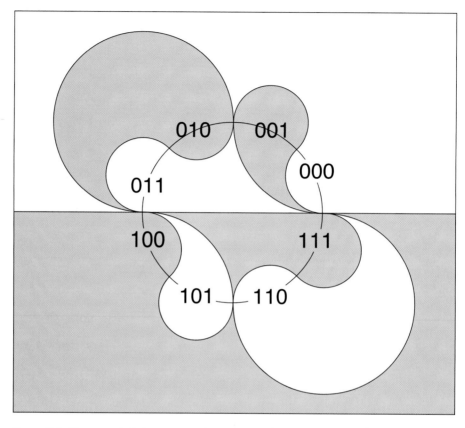

Figure 5.8. The central circle traverses the regions in binary-counting order.

meration presents a real problem: there are hybrid forms with both simple and multiple boundary crossings, as in Figure 5.10. No general results are so far known for the number of Venn diagrams for *n* sets.

As already mentioned, progressing from cosine curves to sine curves is equivalent to a phase shift of the curves, and in the circular form of the diagram this can be achieved by simple rotation. In consequence it is easy to design a kind of circular slide rule which can change from one form to the other, such as I presented at the

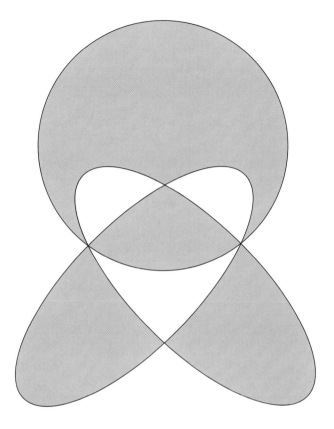

Figure 5.9. Three-set binary-form Venn diagram (Edwards and Smith 1989).

1989 meeting of the International Statistical Institute in Paris. Instructions for making it are given in Appendix 2.

NOTES

Professor C. A. B. Smith died in January 2002. For an obituary see Edwards (2002).

Cosine Curves and Sine Curves

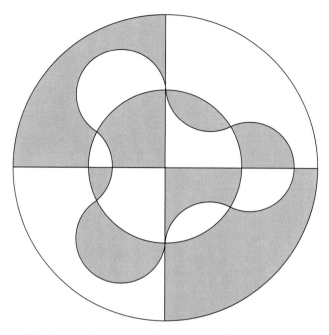

Figure 5.10. One of the four-set forms intermediate between the Gray code and binary forms.

CHAPTER 6

Ironing
the Hypercube

What exactly *is* a Venn diagram? What problem does it solve? What is it really trying to tell us?

Well, solving problems in the propositional calculus is no longer a very lively field of research, but computer science is, and it is here that the fundamental nature of the diagram becomes clear. *The dual of a Venn diagram is a maximal planar subgraph of a Boolean cube.* In this chapter I explain this statement and show how it justifies the assertion that a Venn diagram is trying to tell us as much about a cube in n dimensions as we can represent on a piece of paper. It is a two-dimensional map of an important object in many-dimensioned space, the *Boolean cube.*

In previous chapters we have seen how one way of looking at a Venn diagram is as a map of the binary numbers, an n-set diagram with regions corresponding

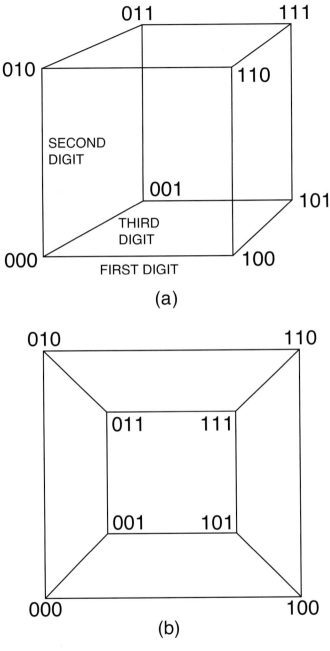

Figure 6.1. Two views of a Boolean cube in three dimensions, showing how (a) the cube can be drawn as (b) a planar graph.

to the numbers 0 to 2^n-1. That was a convenient way to explain a Venn diagram, but if we were to set out to create a representation of the binary numbers we would not naturally think of starting with such a diagram. We would be more likely to start with a Cartesian representation, using a dimension for each digit. With two digits a simple square appears, and with three, a cube (Figure 6.1a). Already we are lumbered with a three-dimensional figure which on our two-dimensional paper we can only draw in perspective, though in the case of a cube we can in fact capture its essence by a slight rearrangement, as in Figure 6.1b. In this view none of the sides intersect each other, which leads graph theorists to describe it as a *planar graph.*

We have now arrived at a map of the binary numbers 000 to 111 with the interesting property that two points (the numbers are represented by points) are joined by a line if and only if the two numbers differ in just one digit. But isn't that what a Venn diagram does? After all, Figure 1.3b shows a labeled Venn diagram, and its adjacent regions (representing numbers) differ in just one digit. Figure 6.2a shows the same thing in Edwards–Venn form (see Figure 3.6a), similarly labeled. Let us now represent each region by a point, and each boundary between adjoining regions by a line joining the respective points (Figure 6.2b). Figure 6.1b, the planar graph of a cube, has now reappeared. Every region in the diagram corresponds to a point in the graph, and every pair of regions with a common boundary corresponds to a pair of points joined by a line. Thus with three digits (and three dimensions) all the edges of the Boolean cube have corresponding edges in the Venn dual graph.

However many digits are being considered, the dual graph is evidently always *planar*—no lines cross. But with more than three digits it might be only a *subgraph*, that is, only part of the Boolean cube, because there is no guarantee in the construction of a Venn diagram that *every* pair of regions representing numbers which

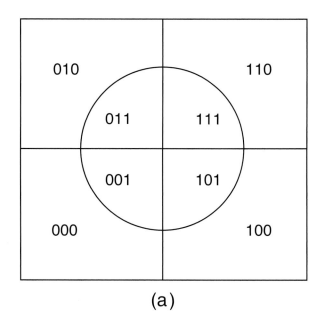

(a)

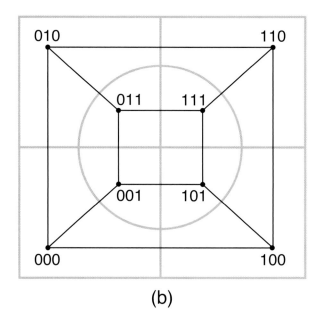

(b)

Figure 6.2. *(a)* Three-set Edwards–Venn diagram and *(b)* its dual graph. Compare Figure 6.1.

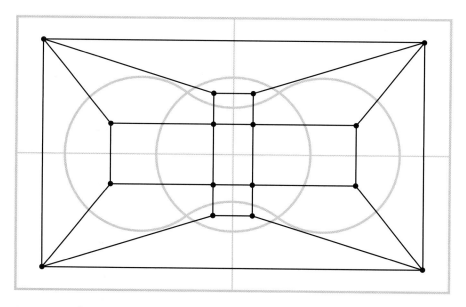

Figure 6.3. The dual graph of a four-set Venn diagram.

differ only in a single digit are adjacent, although the comparable property is intrin-
sic to a Boolean cube.

We cannot so easily draw a four-digit Boolean cube, but it certainly has thirty-
two edges ($n \cdot 2^{n-1}$ for n digits), whilst the dual graph of a four-set Venn diagram
(Figure 6.3) has only twenty-eight edges. Four edges are missing, so it can only be a
subgraph, but it is certainly maximal. In other words, if you want to "iron out" a
four-digit Boolean cube onto a piece of paper so that none of its edges cross you will
have to remove just four of its edges. In general, with n digits the number of edges
in the dual Venn graph is $2^{n+1} - 4$, $n > 1$, so that the deficiency increases quite rapidly with the number of digits (Table 6.1). Nevertheless this is the best that can be
done, and although one can think of other ways of generating the maximal planar
subgraph, the dual graph of an Edwards–Venn diagram is the most direct and intu-

Ironing the Hypercube

Table 6.1.

The number of edges in an n-cube (a)

and in the dual graph of an n-set Venn diagram (b)

n	1	2	3	4	5	6
(a)	1	4	12	32	80	192
(b)	1	4	12	28	60	124

itively obvious way. Note that we can immediately identify in Figure 6.3 which points the missing edges would be linking, for they must be the ones with fewer than four incident edges, and there are indeed eight points with only three incident edges.

This "maximal" sense makes Venn diagrams doubly impressive. It is not just that they possess the properties required of them in the first place, but that they turn out to do so in a manner which displays as much further information as possible, given the constraints of their two-dimensional format. Computer scientists use *Karnaugh maps* and other similar representations for much the same purpose, and in the future a balance will no doubt be reached amongst the various diagrams, each of which has its own strengths and weaknesses. We may also note that the *Hamming codes* in information theory may be described using Venn diagrams and (lest we be thought ignorant of it) that of course the Gray code listing of the vertices of a Boolean cube is a *Hamilton circuit* on the cube, so that discovering Hamilton circuits is the same as discovering Gray codes. But if we investigate their graph-theoretical and com-puter-science properties too deeply we risk too great a departure from our original objective of exploring Venn diagrams as visual geometrical objects, so in the next chapter we return to the visual arts by introducing perhaps the most striking Venn diagram of all.

NOTES

I first realized that the dual of a Venn diagram was a maximal planar subgraph of a Boolean cube in 1990, but the paper noting this was rejected by two professional journals (but see my popular article "How to Iron a Hypercube" [1991a], where I commented that Venn diagrams and their dual graphs "are the best way of representing as much of the information contained in a hypercube as can be put onto the plane without any spurious intersections"). Dr. Charles Little of the Department of Mathematics, Massey University, New Zealand, showed me the simple proof in January 1996. The earliest I have seen the observation in print is in a paper by Chilakamarri, Hamburger, and Pippert (1996a), who gave the proof which, although trivially short, assumes a knowledge of graph theory and is therefore omitted here. I similarly omit the simple demonstrations of the formulae behind the figures in Table 6.1.

For other results connecting Venn diagrams and graph theory see Chilakamarri, Hamburger, and Pippert 1996b. Winkler (1984) observed that the dual graph is a subgraph of a Boolean cube and that if and only if it is Hamiltonian (i.e., contains a Hamilton circuit) can another set be added to the Venn diagram. For further remarks on extending and enumerating Venn diagrams, see the end of chapter 7.

Diagrams with Rotational Symmetry

I mentioned in chapter 3 that Branko Grünbaum, the doyen of Venn diagram explorers, in 1975 published a diagram using only convex sets which was in principle capable of extension to any number of sets, but that it lacked the visual appeal of an Edwards–Venn diagram (although it shared the same "Lewis Carroll" symmetry).

However, another diagram in the same paper combined beauty and novelty to an astonishing degree: Grünbaum had discovered how to construct a Venn diagram with five congruent ellipses, and in so doing had almost casually demonstrated the existence of a new kind of Venn diagram, as well as showing that five-fold rotational symmetry is attainable (Figure 7.1). In due course it was this rotational symmetry that fired my own imagination and led to further discover-

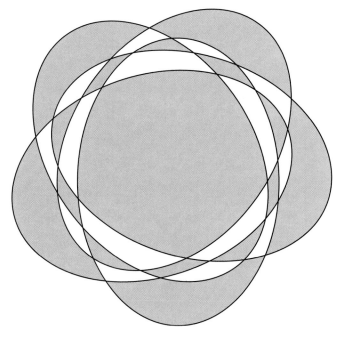

Figure 7.1. Grünbaum's 1975 irreducible five-set diagram (redrawn and shaded).

ies, as this chapter relates. But first we must study what new kind of diagram Grünbaum had invented.

In chapter 5 we looked at Venn diagrams which were not *simple,* in that the set boundaries were no longer constrained to intersect only in pairs, but we still assumed that each was obtained by adding an nth set to a diagram for $n-1$ sets. Such diagrams are said to be *reducible,* for it is always possible to reduce the number of sets by one and still be left with a Venn diagram. One of the many remarkable things about Grünbaum's five-set diagram is that it is *irreducible:* take away any ellipse and we readily see that one of the others then has *two* of its edges exposed to the outside, so that the corresponding regions both represent the same thing.

In 1880 Venn wrote, "With five terms ellipses fail, at least in the above simple form,"

Diagrams with Rotational Symmetry

a statement he repeated in both editions of *Symbolic Logic*. The qualification is of some importance, however, for the "simple form" referred to was his four-set diagram, which he was trying to extend. He had no notion whatsoever that for more than four sets there might be such things as irreducible diagrams which nevertheless possessed all the properties deemed necessary for a logic diagram. Grünbaum's five-set diagram might therefore be held not to invalidate Venn's statement. But in 2000 Hamburger and Pippert did manage to construct an example of a simple reducible five-set diagram, using five congruent ellipses—just (some of the regions are mere slivers).

It was the symmetry of Grünbaum's diagram, however, which came to fascinate me. Since 1988, when I first learned of his work, I had been sending him my material but without any response—until November 1992, when I was visiting a former student of mine, Professor Elizabeth Thompson, who worked in the same building at the University of Washington. I was able to have a few words with Professor Grünbaum, and as we parted he gave me a copy of his "Venn Diagrams I," recently published in the first volume of a new journal, *Geombinatorics* (Grünbaum 1992a).

An *n*-set diagram is said to be *symmetrical* when a rotation of $2\pi/n$ radians about some point (its *center*) leaves the diagram unchanged ("*n*-fold symmetry"). Note that as well as Grünbaum's five-set, Venn's original three-set is symmetrical, though a simple four-set cannot be drawn with this kind of symmetry. On a flight from Seattle to Los Angeles I read Grünbaum's paper and was fascinated by the fivefold symmetry of his beautiful diagram (my Figure 7.1), on which he did not, however, comment. From Los Angeles I took a fifteen-hour flight to Sydney, during which I came to realize that just as the Edwards–Venn form of a diagram exhibited "necklaces" of regions which correspond to all the selections of just *r* things from *n* different things (see chapter 4 and Figure 4.7), so might this be true of Grünbaum's symmetrical diagram—and indeed it was. Moreover, in that symmetrical form the region interior to all the sets (corresponding to the selection of all *n* things, or to the number 11111) is the central one, so the successive necklaces are themselves rotationally symmetrical (see Figure 7.1).

Not surprisingly for someone who had written a book entitled *Pascal's Arithmetical Triangle* (1987, 2002), I knew Leibniz's theorem, that every binomial coefficient in the *n*-row of Pascal's triangle except the two terminal 1s is divisible by *n* if and only if *n* is prime. Thus in the 4-row 1 4 6 4 1 we find 6, which is not divisible by 4, but in the 1 5 10 10 5 1 row we find 10, divisible by 5, and so on. Thinking about the necklaces I realized that if *n,* the number of sets, was prime, each set might possess *n*-fold symmetry, and if they all did then so might the diagram as a whole. Not only was this true for Grünbaum's "5" but obviously it was true for Venn's "3" as well.

These two cases in fact exhibit an even greater symmetry, which we may refer to as *polar.* Recall that a natural home for a Venn diagram is the surface of a sphere, and this is particularly true of diagrams that exhibit rotational symmetry. Venn's three-set diagram not only has threefold symmetry if the pole of projection is taken to be the center of an octant (as in Figure 2.7*a*), but it also has *polar symmetry,* because the two hemispheres are congruent. I call diagrams with both rotational and polar symmetry *completely symmetrical.*

On 15 November 1992, in Adelaide, South Australia, I started to look for a completely symmetrical seven-set diagram, fortunately unaware that anyone else had ever attempted such a challenge, let alone succeeded in it. The following weekend I wrote an account of how I constructed one. It conveys something of the excitement of the chase:

I drew a heptagon, surrounded it with 7 four-sided regions, then surrounded the resulting figure with 21 little regions, and had a shot at seeing where the 35 regions next required could be fitted in. It all ran into a muddle, but there was hope.

On Monday evening, 16 November, I was bitten by a dog, so on the Tuesday I stayed in [college] in the morning. I abandoned the direct "circular" approach to constructing the 7-set diagram and started trying to make a linear diagram which was clearer to work with. Two attempts failed because it turned out that though I had arranged the correct numbers of regions 7, 21, 35, etc., two of the sets did not intersect at all, so some of the regions were repeats. So I tried 5 sets in the same format so as to get the hang of it, and that was easy. As soon as I rearranged the result in circular format the possibility of ellipses became obvious—one just needs a non-circular curve to go through four points, and an ellipse is the obvious choice.

So I returned to the case of 7 with renewed vigour. I cut the second of the earlier diagrams along the mid-line and slid the parts relative to each other until I found a position where, on analysis, I had seven sets all the same shape [Figure 7.2]. I eagerly and rap-

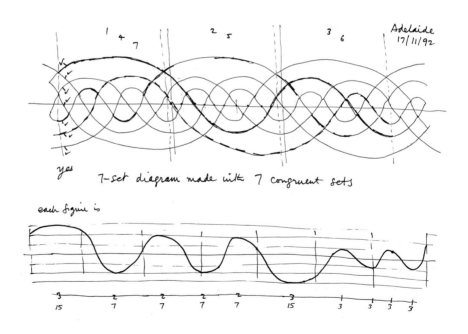

Figure 7.2. Seven-set diagram composed of seven congruent sets.

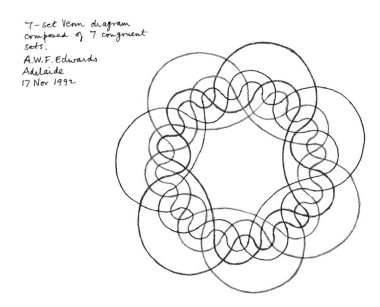

7-set Venn diagram composed of 7 congruent sets.
A.W.F. Edwards
Adelaide
17 Nov 1992

Figure 7.3. Seven-set Venn diagram composed of seven congruent sets.

idly rearranged what I had discovered in Rose Window form to produce the elegant result of the figure [Figure 7.3], which I firmly dated 17 November 1992.

The following weekend I wrote the corresponding computer program, which generated such a beautiful design I could hardly bear to turn it off [Figure 7.4]. I also made a 5-set by the same method, which is in some ways even more beautiful than Grünbaum's ellipses [Figure 7.5].

I immediately sent my work to Grünbaum in Seattle, and settled down to rereading some of the papers to which his "Venn Diagrams I" referred, including his 1975 paper. There I noticed his remarks about a 1963 paper by Henderson in which the author had made the very same point about the divisibility of the binomial coefficients leading to the possibility of symmetrical diagrams. Grünbaum wrote, "Henderson men-

Figure 7.4. The completely symmetrical seven-set diagram "Adelaide."

tions that he has found a symmetric Venn diagram of 7 hexagons. The present author's search for such a diagram has been unsuccessful, as have attempts to clarify Henderson's claim; at present it seems likely that no such diagram exists."

I was naturally pleased to read this, even though it roused the suspicion that Henderson's idea of rotational symmetry might have been planted in my mind if I had first read his paper in 1988 when searching the literature for precursors of the Edwards–Venn diagram. Consulting my notes of 1988 I find the following comment on Henderson (1963): "Does 5 classes, and says 7 can be done, but he is concerned only to make rotational-symmetric diagrams which employ identical figures for each set. No help." Isn't the human mind wonderful? Because Henderson's paper did not contain what I was looking for at the time, I merely buried it in my notes with a dismissive comment on a topic I was to find of absorbing interest four years later!

Figure 7.5. Three completely symmetrical Venn diagrams.

Most deep understanding of other people's work comes from having unwittingly repeated it oneself, for only *then* does one see what the others meant all along. In fact Henderson did offer the first two five-set symmetrical diagrams, but neither was simple; Grünbaum's was the first simple one.

Conscious of Waring's Rule of 1773—"The person who first publishes an idea, or communicates it to his friends, deserves to be called its inventor"—I sent my diagram to the *Adelaidean,* the newspaper of the University of Adelaide (where I was on sabbatical leave from Cambridge). The paper published it on 14 December 1992, by which time I had left Adelaide for Hamilton, New Zealand. There, on 7 December, Professor Elizabeth Thompson from Seattle, who was attending the same meeting in Hamilton, passed me a copy from Professor Grünbaum of his paper "Venn Diagrams II" (Grünbaum 1992b). This paper, in the issue of *Geombinatorics* dated

Diagrams with Rotational Symmetry

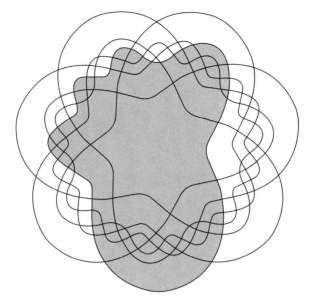

Figure 7.6. The completely symmetrical seven-set diagram "Hamilton," showing the shape of a single set.

October 1992, contains the first two seven-set symmetrical diagrams to be published. Neither of them possesses polar symmetry. Correspondence between us later established that an unpublished diagram drawn by Grünbaum early in 1992 was isomorphic to my "Adelaide." Whilst in Hamilton I found another, on 8 December, and christened it "Hamilton." Figure 7.6 shows a version of "Hamilton" with its "footprint"—the shape of each set—indicated.

I then embarked on writing a computer program designed to enumerate exhaustively the whole family of completely symmetrical seven-set diagrams constructed on the "necklace" principle. It turned out that there were six (Figure 7.7), the four new ones being named "Massey," "Palmerston North," and "Manawatū" (New Zealand) and "Victoria" (British Columbia) for reasons which are explained elsewhere (Edwards 1998). My 1998 paper relates how "Victoria" initially escaped notice through my inad-

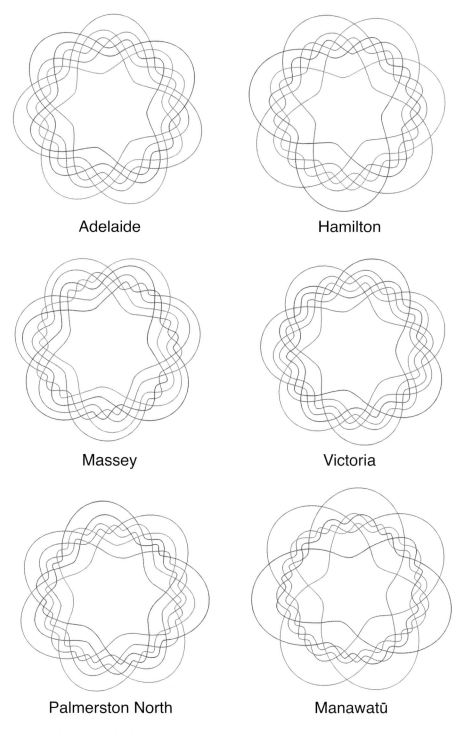

Adelaide

Hamilton

Massey

Victoria

Palmerston North

Manawatū

Figure 7.7. The six completely symmetrical seven-set diagrams constructed on the "necklace" principle.

vertently omitting it from a hand-compiled list based on the computer output, only to be reinstated (and named) by Professor Frank Ruskey of the University of Victoria.

The enumeration of all the simple seven-set symmetrical diagrams, with and without polar symmetry and with and without the "necklace" structure, is a formidable task, and greater interest naturally attaches to the construction of a completely symmetrical simple diagram for eleven sets, the next prime number. I thought I had constructed one in 1993 (see reference 2 in Grünbaum 1999), but only years later did I write the computer program needed to check that each binary number from 0 to $2^{11} - 1$ corresponded to one, and only one, region of the diagram—which turned out not to be the case. Hamburger (2002) has succeeded in constructing a non-simple symmetrical eleven-set diagram, and early in 2003 it was reported that an existence proof for such diagrams has been obtained for all primes.

NOTES

Grünbaum (1999) calls a diagram with polar symmetry *self-complementary,* because such a diagram is isomorphic to the one found by exchanging 0s and 1s in the binary labeling of the regions. This is preferable for diagrams in the plane, but I always have at the back of my mind the spherical version. Readers curious about the origin of Waring's Rule can consult Edwards 1991b. In my 1998 paper I mention the discovery in 1992 of some additional seven-set symmetrical diagrams lacking polar symmetry; "Hamilton" was published in Edwards 1994.

The existence proof for symmetrical diagrams for all primes was presented at a workshop and reported in *Science* (299 [31 January 2003], 651).

The discovery and enumeration of Venn diagrams of one kind and another will provide work for graph-theorists and combinatorialists for many years to come; the field is quite active, and (at the time of writing) it can be followed in Frank Ruskey's colorful survey in *The Electronic Journal of Combinatorics* (2002).

Metrical Venn Diagrams

It is quite often suggested that for certain applications, the regions of a Venn diagram might be made proportional in area to the frequencies of the combinations which they represent, thus creating a metrical form of diagram. Experience shows that this is in fact rarely worthwhile and, as we shall see here, Venn himself was opposed to the idea.

There is, however, one example from statistics where it has proved of some value. Readers who are familiar with the concept of a 2 x 2 contingency table might like to be introduced to the way in which the "Lewis Carroll" form of a Venn diagram can be adapted so as to represent associations visually, and the following article explains how this can be done in connection with an example from human genetics. The article is reprinted here with the permission of the editor of the *Annals of Human Genetics* and the junior author (*Annals of Human Genetics* [1992], 56, 71–75).

Metrical Venn diagrams

A. W. F. EDWARDS[1] AND J. H. EDWARDS[2]

[1] *Department of Community Medicine, University of Cambridge, Fenner's, Gresham Road,
Cambridge CB1 2ES*
[2] *Genetics Laboratory, Department of Biochemistry, University of Oxford, South Parks Road,
Oxford OX1 3QU*

SUMMARY

A type of Venn diagram is described which enables the observed frequencies in a $2 \times 2 \times 2$ contingency table to be compared with their expectations on the hypothesis of no associations.

INTRODUCTION

The precise areas of the subdivisions of a Venn diagram do not normally carry any significance, but occasionally they have been chosen so as to represent either the actual or the expected frequencies in the various classes.

Thus in an elementary discussion of Bayes's Theorem one of us (A. W. F. Edwards, 1972) used a square diagram with four subdivisions whose areas corresponded to the numbers in the four cells of a 2×2 contingency table, whilst the other (J. H. Edwards, 1980) subdivided a square by drawing rows and columns whose widths were proportional to the numbers in the corresponding row and column totals of an $m \times n$ contingency table, thus making the area of each rectangular subdivision proportional to the corresponding expectation on the hypothesis of no associations. The advantage of this technique is that when the data are added in the form of dots each randomly placed in its correct rectangle, the approximate strength and significance of any association are immediately revealed (Fig. 1, reproduced from J. H. Edwards, 1980).

In this paper we show that it is possible to draw a diagram like Fig. 1 for the case of a $2 \times 2 \times 2$ table.

THE LEWIS CARROLL DIAGRAM

The graphical equivalent of a $2 \times 2 \times 2$ table is a three-set Venn diagram which, like the table, has $2^3 = 8$ cells (including, of course, the empty set). However, Venn's three-circle form (Venn, 1880, 1881) does not lend itself readily to metrical development, and it is better to start with the layout preferred by Lewis Carroll (Figure 2; Carroll, 1896, see Edwards, 1989). The three factors are C versus c, D versus d, and E versus e; the bottom half of the diagram is C, the right half D, and the central square E. Such a diagram has enough degrees of freedom to support a metrical version in which the areas are proportional to the expected numbers in the eight classes on the hypothesis of no association (either first- or second-order).

If the proportions in the three single factors are p_1, p_2 and p_3, and the square is a unit square, the vertical and horizontal dividers will need to be at $x = p_1$ and $y = p_2$. All that then remains to ensure is that the proportion of each of the four rectangles which the central rectangle includes is p_3, and this will be achieved by imagining lines drawn from the join of the vertical and horizontal dividers to the corners of the unit square and placing each corner of the central

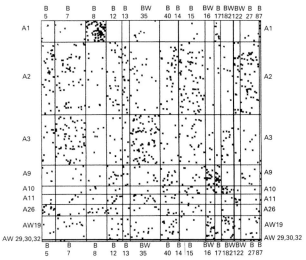

Fig. 1. Showing the distribution of observed HLA-A, HLA-B haplotypes against expectation if not associated. The expectations are represented by the areas within which the points are placed at random (J. H. Edwards, 1980, using data from Tiilikainen, 1978, personal communication).

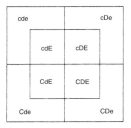

Fig. 2. A Lewis Carroll diagram for three pairs of factors, C-c, D-d and E-e.

Appendix 1

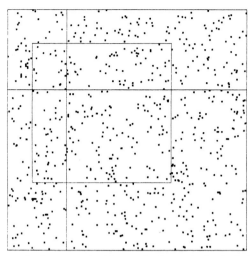

Fig. 3. A diagram for three unassociated factors, showing equal point-density in each area.

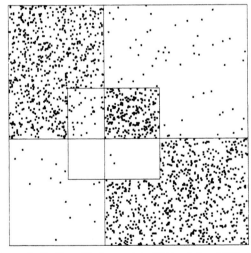

Fig. 4. The diagram for the *Rhesus* blood-group data of Fisher & Race (1946), showing strong associations; the chromosomes are arranged as in Fig. 2.

Table 1. *Estimated rhesus chromosome frequencies (Fisher & Race, 1946) and derived numbers used for the purposes of illustration (Fig. 4)*

CDe	0·4361	809
cde	0·3790	703
cDE	0·1280	237
cDe	0·0305	57
cdE	0·0170	31
Cde	0·0081	15
CDE	0·0013	2
CdE	0·0	0
	1·000	1854

rectangle a fraction $\sqrt{p_3}$ of the way along the corresponding line. The data falling in each cell are then represented, as before, by the correct number of randomly placed dots. A computer program to implement the procedure is easily written.

EXAMPLES: RANDOM AND RHESUS

Figure 3 shows an artificial example in which there are no associations, whilst Fig. 4 shows the *Rhesus* blood-group data of Fisher & Race (1946), where the arrangement of the eight *Rhesus* chromosomes follows the lettering of Fig. 2. Because of the dominance relations the numbers of the different chromosomes cannot be counted but have to be estimated, so the actual data provide rather less information than this representation suggests. To generate counts for the purposes of illustration we have taken the chromosome frequencies as estimated by maximum-likelihood and multiplied them by 2×927, twice the number of blood donors in the sample (Table 1).

The diagram clearly displays what Fisher himself noted, that the chromosomes are not in linkage equilibrium and that they fall into three groups, common, rare and absent (i.e. extremely rare). The extremely rare chromosome, *CdE*, is entirely surrounded by three rare chromosomes, *Cde*, *cdE*, and *CDE*, these being the only three from which it could be derived by the mutation of a single gene, since it is a fundamental property of a Venn diagram that crossing a line changes just a single factor. If *CdE* is deleterious but is maintained by mutation this may explain its rarity, though Fisher actually suggested crossing-over rather than mutation, for which a cube with the 8 chromosomes as its corners is the appropriate diagram because the 12 genotypes heterozygous at a single locus are represented by its edges, the 12 doubly heterozygous genotypes by its face diagonals, and the 4 triply heterozygous by its internal diagonals (Fisher, 1947). Although a Venn diagram (whether in Lewis Carroll form or not) maps the vertices and edges of a cube perfectly, it does not map the diagonals.

EPILOGUE

John Venn would not have approved of our proposal. In the last part of the last chapter of *Symbolic Logic* (Venn, 1881) he inveighs against attributing significance to the length of lines or the extent of areas in logical diagrams, culminating in the final paragraph:

My own conviction is very decided that all introduction of considerations such as these should be avoided as tending to confound the domains of Logic and Mathematics; of that which is, broadly speaking, qualitative,

99

Appendix 1

and that which is quantitative. The compartments yielded by our diagrams must be regarded solely in the light of being bounded by such and such contours, as lying inside or outside such and such lines. We must abstract entirely from all consideration of their relative magnitude, as we do of their actual shape, and trace no more connection between these facts and the logical extension of the terms which they represent than we do between this logical extension and the size and shape of the letter symbols, A and B and C.

REFERENCES

CARROLL, L. (1896). *Symbolic Logic*. London: Macmillan.

EDWARDS, A. W. F. (1972). *Likelihood*. Cambridge: Cambridge University Press.

EDWARDS, A. W. F. (1989). Venn diagrams for many sets. *New Scientist* **121** (1646), 51–56.

EDWARDS, J. H. (1980). Allelic association in man. In *Population Structure and Genetic Disorders*, (ed. A. W. Eriksson), pp. 239–255. London: Academic Press.

FISHER, R. A. (1947). The Rhesus factor: A study in scientific method. *American Scientist* **35**, 95–103.

FISHER, R. A. & RACE, R. R. (1946) *Rh* gene frequencies in Britain. *Nature* **157**, 48–49.

VENN, J. (1880). On the diagrammatic and mechanical representation of propositions and reasonings. *Philosophical Magazine* (Fifth Series) **10**, 1–18.

VENN, J. (1881). *Symbolic Logic*. London: Macmillan.

A Rotatable
Edwards–Venn Diagram

As mentioned at the end of chapter 5, it is possible to construct a circular form of Venn diagram in which the change from the "Gray code" type to the "binary number" type is achieved by rotation of the sets, this being equivalent to the required phase shift. I presented the following instructions for this at the forty-seventh session of the International Statistical Institute in Paris in 1989.

Preparation

1. Photocopy Figures A.2 to A.5 onto overhead projector sheets and cut them out, taking special care with the "ears." The lines on each figure may be colored differently at this stage if you wish; each corresponds to a set boundary.

2. Photocopy Figure A.1 onto white paper and cut it out. The paper should be as thick as possible, or the copied figure should be mounted on card. If you are coloring the set boundaries, the circle and the line should be given different colors since they are the boundaries of different sets.

3. Mount the disks concentrically in the order 1 (the base) to 5 using a drawing pin (thumbtack). A more permanent arrangement is to cut out the little square at the center of each disk and use a plastic popper (snap fastener) or a brass fastener to hold the disks together.

Operation

Figure A.6 shows the appearance of the finished product. When the sets are rotated so that the black ears coincide the result is the "Gray code" form of the diagram. On rotating the sets so that the white ears coincide the "binary number" version with five sets appears.

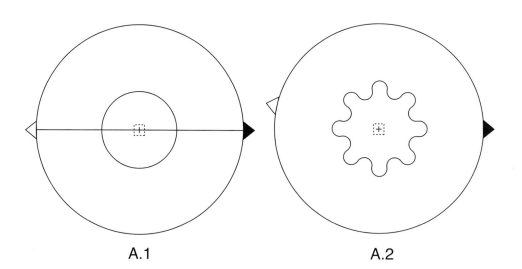

A.1   A.2

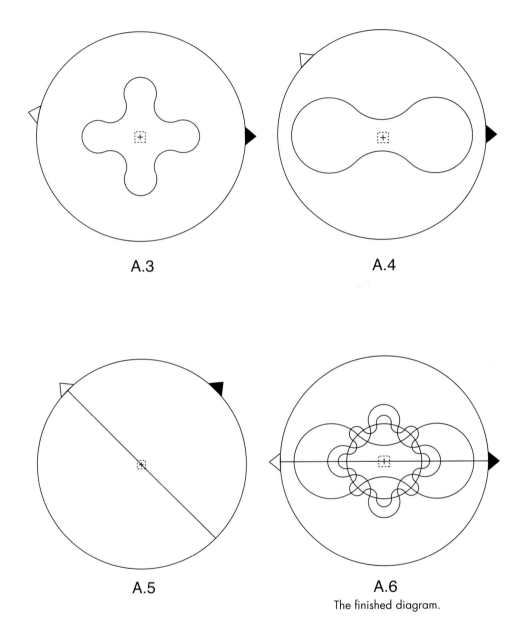

A.3

A.4

A.5

A.6
The finished diagram.

REFERENCES

Allen, A. H. 1871. Lecture experiments on colour. *Nature* 4, 346.

Anderson, D. E., and Cleaver, F. L. 1965. Venn-type diagrams for arguments of N terms. *Journal of Symbolic Logic* 30, 113–18.

Boole, G. 1854. *An Investigation of the Laws of Thought on Which Are Founded the Mathematical Theories of Logic and Probabilities.* London: Walton and Maberly.

Boyd, A. V. 1985. Venn diagram of rectangles. *Mathematics Magazine* 58, 251.

Carroll, L. 1887. *The Game of Logic.* London: Macmillan.

———. 1896. *Symbolic Logic.* London: Macmillan.

———. 1977. *Lewis Carroll's Symbolic Logic,* ed. W. W. Bartley III. Hassocks, Sussex, Engl.: Harvester Press.

Chilakamarri, K. B., Hamburger, P., and Pippert, R. E. 1996a. Venn diagrams and planar graphs. *Geometriae Dedicata* 62, 73–91.

———. 1996b. Hamilton cycles in planar graphs and Venn diagrams. *Journal of Combinatorial Theory, B* 67, 296–303.

Cromwell, P., Beltrami, E., and Rampichini, M. 1998. The Borromean rings. *Mathematical Intelligencer* 20, 53–62.

Didron, A. N. 1843. *Iconographie Chrétienne: Histoire de Dieu.* Paris: Imprimerie Royale.

———. 1886. *Christian Iconography: The History of Christian Art in the Middle Ages.* London: Bell.

Dunham, W. 1999. *Euler, the Master of Us All.* Dolciani Mathematical Expositions 22. Washington, D.C.: The Mathematical Association of America.

106   Edwards, A. W. F. 1987. *Pascal's Arithmetical Triangle.* London: Griffin; 2d ed., 2002, Baltimore: Johns Hopkins University Press.

———. 1989. Venn diagrams for many sets. *New Scientist* 121, no. 1646 (7 January), 51–56.

———. 1991a. How to iron a hypercube. *Mathematical Gazette* 75, 433–36.

———. 1991b. A note on Waring's Rule. *Historia Mathematica* 18, 177.

———. 1993. Mendel, Galton, Fisher. Second Sir Ronald Fisher Lecture, University of Adelaide, 12 November 1992. *Australian Journal of Statistics* 35, 129–40.

———. 1994. The sevenfold symmetric Venn diagrams. In *Proceedings of the XVII International Biometric Conference, Hamilton, Ontario,* 2, 238. Hamilton, Ont.

———. 1998. Seven-set Venn diagrams with rotational and polar symmetry. *Combinatorics, Probability, and Computing* 7, 149–52.

———. 2002. Professor C. A. B. Smith, 1917–2002. *The Statistician* 51, 404–5.

Edwards, A. W. F., and Edwards, J. H. 1992. Metrical Venn diagrams. *Annals of Human Genetics* 56, 71–75.

Edwards, A. W. F., and Smith, C. A. B. 1989. New 3-set Venn diagram. *Nature* 339, 263.

Fauvel, J., Flood, R., and Wilson, R. (eds.). 2000. *Oxford Figures.* Oxford: Oxford University Press.

Fisher, J. C., Koh, E. L., and Grünbaum, B. 1988. Diagrams Venn and how. *Mathematics Magazine* 61, 36-40.

Fisher, R. A. 1935. *The Design of Experiments.* Edinburgh: Oliver and Boyd.

Gardner, M. 1983. *Logic Machines and Diagrams.* Brighton, Engl.: Harvester Press.

Gilbert, M. 1988. *Winston S. Churchill, Volume VIII, 1945–1965, "Never Despair."* London: Heinemann.

Grünbaum, B. 1975. Venn diagrams and independent families of sets. *Mathematics Magazine* 48, 12–22.

———. 1992a. Venn Diagrams I. *Geombinatorics* 1, 5–12.

———. 1992b. Venn Diagrams II. *Geombinatorics* 2, 25–32.

———. 1999. The search for symmetric Venn diagrams. *Geombinatorics* 7, 104–9.

Hamburger, P. 2002. Doodles and doilies: non-simple symmetric Venn diagrams. *Discrete Mathematics* (Special Issue in Honor of the 65th Birthday of Daniel J. Kleitman) 257, no. 2–3, 423–39.

Hamburger, P., and Pippert, R. E. 1996. Simple, reducible Venn diagrams on five curves and Hamiltonian cycles. *Geometriae Dedicata* 68, 245–62.

———. 2000. Venn said it couldn't be done. *Mathematics Magazine* 73, 105–10.

Hammer, E. M. 1995. *Logic and Visual Information.* Stanford, Calif.: CSLI Publications.

Henderson, D. W. 1963. Venn diagrams for more than four classes. *American Mathematical Monthly* 70, 424–26.

Humphries, M. 1987. Venn diagrams using convex sets. *Mathematical Gazette* 71, 59.

Jevons, W. S. 1870. On the mechanical performance of logical inference. *Philosophical Transactions of the Royal Society of London* 160, 497–518.

———. 1877. *The Principles of Science: A Treatise on Logic and Scientific Method,* 2d ed. London: Macmillan.

MacHale, D. 1985. *George Boole: His Life and Work.* Dublin: Boole Press.

Mich, J. 1871. *Grundriss der Logik.* Troppau [Czech Rep.]: Buchholz and Diebel.

More Jr., T. 1959. On the construction of Venn diagrams. *Journal of Symbolic Logic* 24, 303–4.

Obituary of John Venn. 1926. *Proceedings of the Royal Society London A* 110, x–xi.

Poythress, V. S., and Sun, H. S. 1972. A method to construct convex, connected Venn diagrams for any finite number of sets. *Pentagon,* spring 1972, 80–82.

Ruskey, F. 2001. A survey of Venn diagrams. *The Electronic Journal of Combinatorics,* March 15. www.combinatorics.org/Surveys/ds5/VennEJC.html.

Stewart, I. 1989. Les dentelures de l'esprit. *Pour la Science* 138 (April), 104–9.

———. 1992. *Another Fine Math You've Got Me Into. . . .* New York: W. H. Freeman.

Tait, P. G. 1877a. On knots. *Transactions of the Royal Society of Edinburgh* 28, 145–90.

———. 1877b. Some elementary properties of closed plane curves. *Messenger of Mathematics* 6, 132–33.

Venn, J. 1866. *The Logic of Chance.* Cambridge: Macmillan.

———. 1880. On the diagrammatic and mechanical representation of propositions and reasonings. *London, Edinburgh, and Dublin Philosophical Magazine and Journal of Science [Fifth Series]* 9, 1–18.

———. 1881. *Symbolic Logic.* London: Macmillan.

———. 1889. *The Principles of Empirical or Inductive Logic.* London: Macmillan.

———. 1897. *Biographical History of Gonville and Caius College, 1349–1897,* I (1897), II (1898), III (1901). Cambridge: Cambridge University Press.

References

108   Weaver, J. R. H. (ed.). 1937. *Dictionary of National Biography (1922–1930)*. London: Oxford University Press.

Weiss, J. 1997. The Logic Demonstrator of the 3rd Earl Stanhope (1753–1816). *Annals of Science* 54, 375–95.

Wilson, R. 2002. *Four Colours Suffice*. London: Allen Lane.

Winkler, P. 1984. Venn diagrams: some observations and an open problem. *Congressus Numerantium* 45, 267–74.

INDEX

√